ARCHITECTS, CONTRACTORS & ENGINEERS
GUIDE TO CONSTRUCTION COSTS

Section

DIVISION 1 GENERAL REQUIREMENTS 1
    1A Equipment

DIVISION 2 SITE WORK & DEMOLITION 2
    2A Quick Estimating

DIVISION 3 CONCRETE 3
    3A Quick Estimating

DIVISION 4 MASONRY 4
    4A Quick Estimating

DIVISION 5 METALS 5
    5A Quick Estimating

DIVISION 6 WOOD & PLASTICS 6
    6A Quick Estimating

DIVISION 7 THERMAL & MOISTURE PROTECTION 7
    7A Quick Estimating

DIVISION 8 DOORS & WINDOWS 8
    8A Quick Estimating

DIVISION 9 FINISHES 9

DIVISION 10 SPECIALTIES 10

DIVISION 11 EQUIPMENT 11

DIVISION 12 FURNISHINGS 12

DIVISION 13 SPECIAL CONSTRUCTION 13

DIVISION 14 CONVEYING SYSTEMS 14

DIVISION 15 MECHANICAL 15

DIVISION 16 ELECTRICAL 16

SQUARE FOOT COSTS ..... SF

STATISTICAL DATA ..... S

U.S. MEASUREMENTS, METRIC EQUIV'S, & AC&E ANNUAL INDEX ..... MM

ALPHABETICAL INDEX ..... AI

EDITOR'S NOTE
2002

This annually published book is designed to give a uniform estimating and cost control system to the General Building Contractor. It contains a complete system to be used with or without computers. It also contains Quick Estimating sections for preliminary conceptual budget estimates by Architects, Engineers and Contractors. Square Foot Estimating is also included for preliminary estimates.

The Metropolitan Area concept is also used and gives the cost modifiers to use for the variations between Metropolitan Areas whose populations are in excess of 500,000 people. This encompasses over 50% of the industry. This book is published annually on the first of August to be historically accurate with the traditional May-July wage contract settlements and to be a true construction year estimating and cost guide.

The Rate of Inflation in the Construction Industry in 2000 was *3.0%*. Labor contributed a *3.3%* increase and materials rose *2.8%*.

The Wage Rate for Skilled Trades increased an average of *3%* in 2001. Wage rates will probably increase at a *4%* average next year.

The Material Rate increased *2.8%* in 2000. The main increases were in pipe, asphalt and copper products. The major decreases were in lumber and plywood.

Construction Volume should increase in 2001. Housing will probably rise slightly, and Industrial and Commercial Construction should drop slightly. Highway and Heavy Construction should increase.

The Construction Industry again kept inflation under control in 2001. Some Materials should inflate at a slow pace, and Labor should again inflate slightly in 2002.

We are recommending using a *3%* increase in your estimates for work beyond July 1, 2002.

Don Roth
Editor

Metropolitan Area Construction Cost Modifiers and Population (2000 Census)
in the
50 largest Metropolitan Statistical Areas

Metropolitan Area	Population	Rank	Modifier	Metropolitan Area	Population	Rank	Modifier
1. Atlanta	4,112,198	11	90	26. Minneapolis, St. Paul	2,968,806	15	106
2. Austin	1,249,763	37	86	27. Nashville	1,231,311	38	91
3. Boston	5,819,100	7	118	28. New Orleans	1,337,726	34	89
4. Buffalo	1,170,111	42	114	29. New York	21,199,865	1	145
5. Charlotte	1,499,293	33	82	30. Norfolk	1,569,541	30	90
6. Chicago	9,157,540	3	114	31. Oklahoma City	1,083,346	48	89
7. Cincinnati	1,979,202	23	98	32. Orlando	1,644,561	27	94
8. Cleveland	2,945,831	16	110	33. Philadelphia	6,188,463	6	117
9. Columbus	1,540,157	32	100	34. Phoenix	3,251,876	14	98
10. Dallas, Ft Worth	5,221,801	9	85	35. Pittsburgh	2,358,695	21	107
11. Denver	2,581,506	19	96	36. Portland, OR	2,265,223	22	108
12. Detroit	5,456,428	8	114	37. Providence	1,188,613	39	107
13. Grand Rapids	1,088,514	47	100	38. Raleigh, Durham	1,187,941	40	82
14. Greensboro	1,251,509	36	82	39. Richmond	996,512	50	89
15. Hartford	1,183,110	41	108	40. Rochester, NY	1,098,201	46	110
16. Houston	4,669,571	10	87	41. Sacramento	1,796,857	24	116
17. Indianapolis	1,607,486	28	102	42. Salt Lake City	1,333,914	35	86
18. Jacksonville	1,100,491	45	91	43. San Antonio	1,592,383	29	87
19. Kansas City	1,776,062	25	100	44. San Diego	2,813,833	17	112
20. Las Vegas	1,563,282	31	108	45. San Francisco	7,039,362	5	116
21. Los Angeles	16,373,645	2	112	46. Seattle, Tacoma	3,554,760	13	108
22. Louisville	1,025,598	49	92	47. St. Louis	2,603,607	18	108
23. Memphis	1,135,614	43	94	48. Tampa, St. Pete.	2,395,997	20	96
24. Miami	3,876,380	12	88	49. Washington, Balt.	7,608,070	4	94
25. Milwaukee	1,689,572	26	106	50. West Palm Beach	1,131,184	44	93

(1) From Department of Commerce, 2000 Census
(2) AC&E Construction Cost Modifier - No Land Value Included

HOW TO USE THIS BOOK

Labor Unit Columns
- Units *include* all taxable (vacation) and untaxable (welfare and pension) fringe benefits.
- Units *do not include* taxes and insurance on labor (approximately 35% added to labor cost). (Workers Compensation, Unemployment Compensation, and Employers FICA.)
- Units *do not include* general conditions and equipment (approx. 10%).
- Units *do not include* contractors' overhead and profit (approx. 10%).
- Units are Union or Government Minimum Area wages.

Material Unit Columns
- Units *do not include* general conditions and equipment (approx. 10%).
- Units *do not include* sales or use taxes (approx. 5%) of material cost.
- Units *do not include* contractors' overhead and profit (approx. 10%).
- Units are mean prices FOB job site.

Cost Unit Columns (Subcontractors)
- Cost is complete with labor, material, taxes, insurance and fees of a subcontractor. This also applies to Quick Estimating sections.
- Units *do not include* general contractors' overhead or profit.

Quick Estimating Sections - For Preliminary and Conceptual Estimating
- Includes all labor, material, general conditions, equipment, taxes, insurance and fees of contractors combined for items not covered in Cost Unit columns.

Construction Modifiers - For Metro Area Cost Variation Adjustments
- Labor items are percentage increases or decreases from 100% for labor units used in the book's labor unit columns. (See first page of each Division.)
- Material and Labor combined are percentage increases and decreases from 100% for metro variations used in the book's Unit Cost columns, Quick Estimating sections and Sq.Ft. Cost Section - see Page 3.
- Recommend Unit Costs of 10% less than Book Units for Residential Construction.

Metric Equivalents & U.S. Weights & Measures Modifiers - See Page 3A-18.

EXAMPLE:	UNIT	LABOR	MATERIAL	COST
2" x 4" Wood Studs & Plates - 8' @ 16" O.C.	BdFt	.60	.52	1.12
Add 35% Taxes & Insurance on Labor		.21	-	.21
Add 5% Material Sales Tax		-	.03	.03
				1.36
Add 10% General Conditions & Equipment				.14
				1.50
Add 10% Overhead & Profit				.15
BdFt Cost - Bid Type Estimating (from page 6-3)			BdFt	1.65
or SqFt Wall Cost w/ 3 Plates - Quick Estimating (page 6A-14)			or SqFt	1.35

See Division 4 for Other Examples (Masonry)

Purchasing Practice (usual) in the industry for each section of work is indicated immediately after section heading in parenthesis and defined as follows:
- **L&M** - Labor and Material Subcontractor - Firm bid to G.C. according to plans and specifications for all labor and material in that section or subsection.
- **M** - Material Subcontractor - Firm bid to G.C. according to plans and specifications for all material in that section or subsection.
- **L** - Labor Subcontractor - Firm bid to G.C. to *install only* material furnished according to plans and specifications by others.
- **S** - Material Supplier - Firm Price for materials delivered only.
- **E** - Equipment Supplier - Firm Price for Equipment Rented or Purchased.

Installation Practice by particular trades are listed behind each section heading and are according to "Agreement and Decisions Rendered Affecting the Building Industry," by the A.F.L. Building and Construction Trades Department, A.F.L.-C.I.O. Area variations, jurisdictional problems and multiple jurisdictions are possible.

		PAGE
0101.0	**GENERAL CONDITIONS - Specified (A.I.A. Doc. A201, 1987)**	**1-2**
.1	SUPERINTENDENT	1-2
.2	PERMITS AND FEES	1-2
.3	TAXES	1-2
.31	Sales and Use Tax	1-2
.32	Unemployment Compensation - State	1-2
.33	Unemployment Compensation	1-2
.34	F.I.C.A. - Employer's Share	1-2
.4	INSURANCE - CONTRACTOR'S LIABILITY	1-2
.5	INSURANCE - HOLD HARMLESS	1-2
.6	INSURANCE - ALL RISK	1-2
.7	CLEANUP	1-2
.8	PERFORMANCE AND PAYMENT BONDS	1-2
.9	SAFETY OF PERSONS AND PROPERTY	1-2
.10	TESTS AND SAMPLES	1-2
0102.0	**SPECIAL CONDITIONS (Changes to G.C. 0101 and Temporary Utilities and Facilities Specified)**	**1-3**
.1	TEMPORARY OFFICE	1-3
.2	TEMPORARY STORAGE SHED	1-3
.3	TEMPORARY SANITARY FACILITIES	1-3
.4	TEMPORARY TELEPHONE	1-3
.5	TEMPORARY POWER AND LIGHT	1-3
.6	TEMPORARY WATER	1-3
.7	TEMPORARY HEAT	1-3
.8	TEMPORARY ENCLOSURES AND COVERS	1-3
.9	TEMPORARY ROADS AND RAMPS	1-3
.10	TEMPORARY DEWATERING	1-3
.11	TEMPORARY SIGNS	1-3
.12	TEMPORARY FENCING, BARRICADES AND WARNING LIGHTS	1-3
0103.0	**MISCELLANEOUS CONDITIONS - Non-Specified**	**1-4**
.1	FIELD ENGINEERING AND LAYOUT	1-4
.2	TIMEKEEPERS AND ACCOUNTANTS	1-4
.3	SECURITY	1-4
.4	OPERATING ENGINEERS FOR HOISTING	1-4
.5	TRAVEL AND SUBSISTENCE - Field Engineers	1-4
.6	TRAVEL AND SUBSISTENCE - Office Employees	1-4
.7	OFFICE SUPPLIES AND EQUIPMENT	1-4
.8	COMMUNICATIONS EQUIPMENT	1-4
.9	HOISTING EQUIPMENT	1-4
.91	Towers and Hoists	1-4
.92	Cranes - Climbing - 20' Override	1-4
.93	Cranes - Mobile Lifting (No Operator)	1-4
.94	Fork Lifts	1-5
.95	Platform Lifts	1-5
.10	TRUCKS AND OTHER JOB VEHICLES (Site Use)	1-5
.11	MAJOR EQUIPMENT	1-5
.12	SMALL TOOLS	1-5
.13	HAULING EQUIPMENT AND TOOLS	1-5
.14	GAS AND OIL FOR EQUIPMENT	1-5
.15	REPAIRS TO EQUIPMENT AND TOOLS	1-5
.16	PHOTOGRAPHS - 2 Views, 4 Copies	1-5
.17	SCHEDULING - C.P.M., Pert, Network, Analysis	1-5
.18	COMPUTER COSTS	1-5
.19	PERFORMANCE BOND - SUBCONTRACTORS	1-5
.20	PUNCH LIST AND FINAL OUT	1-5
1A	**EQUIPMENT**	**1-6A and 1-7A**

DIVISION #1 - GENERAL REQUIREMENTS

General Requirements amount to approximately 6% to 15% of total job costs for General Construction and 5% to 10% for Residential Construction. All items below are included in G.R. except 0101.32, .33, and .34 (taxes and insurance on labor). See Square Foot Costs (SF Section) for more accurate Dollar Value of Job Costs.

0101.0 GENERAL CONDITIONS - Specified (AIA Doc A201 1987)

			UNIT	LABOR	MATERIAL
.1	SUPERINTENDENT (Art. 3.9) Job Size/Complexity		Week	$1,500	-
.2	PERMITS AND FEES (Art. 3.7)				
	Concrete and Commercial	- Approx. Average	MCuFt	-	$2.50
	Residential (Finish Area)	- Approx. Average	SqFt	-	$ .50
	Remodeling	- Approx. Average	M$	-	$5.00
	Varies with Each Permit Division				
.3	TAXES (Art. 3.6)				
.31	Sales and Use Tax (Varies as to State)		%		Avg. 5.5%

		Payroll Max	Per $100 or % Gross Payroll	
.32	Unemployment Compensation	to $ 19,100	State Maximum	9.1%
	(Varies as to State)		State Minimum	.7%
.33	Unemployment Compensation	to $ 7,000	Fed. Maximum	.8%
.34	FICA - Employer's Share	to $ 76,200		6.20%
	Medicare	No Limit		1.45%

.4	INSURANCE - CONTRACTOR'S LIABILITY			
	(Worker's Compensation, P.L. & P.D.) (Art. 11.1.1)			
	Classification Avg - Contractors Experience		Gross Payroll	14.0%
	Clerical - Office	(#8810)	" "	1.0%
	Supervision	(#5606)	" "	3.0%
	Excavation	(#6217)	" "	12.0%
	Concrete - Structural - N.O.C.	(#5213)	" "	15.0%
	Slabs on Ground, Flat	(#5221)	" "	14.0%
	Masonry	(#5022)	" "	11.0%
	Iron and Steel - N.O.C.	(#5057)	" "	32.0%
	Carpentry - N.O.C. Commercial	(#5403)	" "	32.0%
	Interior & Dwellings	(#5437)	" "	16.0%
	Wallboard	(#5445)	" "	13.0%
	Plumbing			7.0%
	Electrical			6.0%
	Wrecking (Demolition)			35.0%
	PL. and PD.			1.5%
	Add for Assigned Risk			75%

			UNIT	LABOR	MATERIAL
.5	INSURANCE - HOLD HARMLESS				
	(By Owner) (Art. 11.2) (Contract Amount)		M$		$ .80
.6	INSURANCE - ALL RISK (Fire/Ext.Cov./Theft/etc.)				
	(By Owner) (Art. 11.2) ($1,000 Deductible)				
	Fire Proof - All Concrete		Annual)		$2.00
	Fire Resistive - Covered Steel		Insurable)		$2.50
	Ordinary - Masonry, Ext. Walls & Joists		Value)		$3.50
	Frame, Wood		per M$)		$6.00
	Deduct for $2,500 Deductible				10%
	Add for Light Protection Areas - Approx.				50%
.7	CLEANUP (Art. 3.15)				
.71	Progress		Bldg - SqFt	$.15	$ .05
.72	Glass, 2 Sides		Glass Area - SqFt	$.30	-
.8	PERFORMANCE & PAYMENT BOND	(If Specified) (Art. 11.4)			
	$ -0- to $ 99,999	Job Bureau Comp. Contr. Amount			2.50%
	$ 100,000 to $ 499,999				1.50%
	$ 500,000 to $2,499,999				1.00%
	$2,500,000 to $4,999,999				.75%
	$5,000,000 to $7,499,999				.70%
	$7,500,000 Up				.65%
.9	SAFETY OF PERSONS & PROPERTY (Art. 10.1 and 10.2) - Job Condition				
.10	TESTS AND SAMPLES (Art. 3.1) - See Appropriate Division				

See 0102.0 Temporary Fences and Barricades

0102.0 SPECIAL CONDITIONS (Changes to G.C. 0101 and Temporary Utilities & Facilities, usually Specified)

		UNIT	LABOR	MATERIAL
.1	TEMPORARY OFFICE			
	Fixed - 10' x 16" Wood	Each	1,300	2,200
	Mobile (Special Constr.) 8' x 32'	Each	900	8,000
		Month	-	380
	Mobile (Special Constr.) 10' x 50'	Each	1,000	12,000
		Month	-	600
	Mobile (Special Constr.) 12' x 60'	Each	1,200	14,000
		Month	-	700
	Add for Delivery & Pickup	Each	-	400
.2	TEMPORARY STORAGE SHED -12' x 24'	Each	1,600	3,200
.3	TEMPORARY SANITARY FACILITIES - Chemical Toilet	Month	-	90
.4	TEMPORARY TELEPHONE (No Long Distance) - Per line	Month	-	80
	TEMPORARY TELEPHONE	Initial	-	250
	TEMPORARY TELEPHONE - Cellular	Month	-	100
	TEMPORARY FAX	Initial	-	1,100
.5	TEMPORARY POWER AND LIGHT			
	Initial Service 110V	Job	-	1,100
	220V	Job	-	1,700
	440V	Job	-	3,200
	Monthly Service and Material	Month	-	220
	Temporary Wiring and Lighting-Total Job	SqFt	-	.07
.6	TEMPORARY WATER -Initial Service	Initial	Job Condition	
	Monthly Service	Month	-	73.00
.7	TEMPORARY HEAT (after Enclosed)			
	See Divisions 3 & 4 for Temporary Heat, Concrete & Masonry			
	See Division 1A for Equipment Rental Rates			
	Heaters:			
	Steam & Unit - Heater Rent - SqFt Htd Area	Month	-	.10
	Piping - SqFt Htd Area	Month	-	.12
	Condensate - SqFt Htd Area	Month	-	.17
	Mobile Home Heater Rent - SqFt Htd Area	Month	-	.09
	Fuel SqFt Htd Area	Month	-	.22
	Tending 8 - Hour/Day	8 Hrs	-	208.00
.8	TEMPORARY ENCLOSURES AND COVERS			
	Per Door - Wood	Each	28.00	36.00
	Per Window - Polyethylene and Frame	Each	10.00	8.00
	Walls - Frame and Reinforced Poly - 1 use	SqFt	.20	.16
	2 uses	SqFt	.18	.12
	Frame and 1/2" Plywood - 1 use	SqFt	.55	.90
	2 uses	SqFt	.45	.55
	Floors- 2" Fiberglass & 7 mil Poly- 1 use	SqFt	.10	.35
.9	TEMPORARY ROADS & RAMPS - No Surfacing	SqYd	Job Condition	
.10	TEMPORARY DEWATERING - No Tending			
	Pumping - 2" Pump	Day	-	87.00
	3" Pump	Day	-	98.00
	4" Pump	Day	-	113.00
	Add for Tending			
	Well Points - LnFt Header	Month	-	97.00
	Add Each Month	Month	-	73.00
.11	TEMPORARY SIGNS - 4' x 8'	Each	100.00	420.00
.12	TEMP. FENCING, BARRICADES & WARNING LIGHTS			
	6' Chain Link - Rented (Erected and Removed)	LnFt	1.10	4.30
	Add for Barbed Wire	LnFt	.30	.78
	Add for Gates	LnFt	7.50	16.00
	Snow Fence - New	LnFt	.37	1.00
	Concrete Barriers (Rented - Month)	LnFt	-	3.25
	Flasher Warning Light (Rented - Month)	Each	-	18.00
	8' Barricade (Rented - Month)	Each	-	40.00
	12' Barricade (Rented - Month)	Each	-	50.00

DIVISION #1 - GENERAL REQUIREMENTS

0103.0	MISCELLANEOUS CONDITIONS (Non-Specified)	UNIT	LABOR*	MATERIAL
.1	FIELD ENGINEERING AND LAYOUT	Week		Job Cond.
.2	TIMEKEEPERS AND ACCOUNTANTS	Week		Job Cond.
.3	SECURITY	Week		Job Cond.
.4	OPERATING ENGINEERS FOR HOISTING	Week		Job Cond.
.5	TRAVEL AND SUBSISTENCE - Field Employees			Job Cond.
.6	TRAVEL AND SUBSISTENCE - Office Employees			Job Cond.
.7	OFFICE SUPPLIES AND EQUIPMENT	Week		.03%
.8	COMMUNICATIONS EQUIPMENT			
	Radio-Multi Channel Hand Unit w/ Charger & Battery	Each	-	1,250
	Base Unit - F.M. with Antenna and Mike	Each	-	1,400
.9	HOISTING EQUIPMENT (Towers, Cranes and Lifts)			
.91	Towers and Hoists - Portable Self-Erecting Style			
	Skip (Based on 100' Unit) -3000# to 5000# Capacity	Month	-	2,000
	2000# to 3000# Capacity	Month	-	1,900
	Add to Heights Above 100'	LnFtMo	-	20
	Towers - Fixed (based on 100') - 30' Override			
	Tubular Frame - Double Well - 5000# Capacity	Month	-	1,750
	Installation - Double Well - Subcont - Up	LnFt	55	-
	Double Well Down	LnFt	40	-
	Add on or Deduct from 100' - Month	LnFt	-	20
	Add for Concrete Bucket and Hopper	Month	-	380
	Add for Chicago Boom	Month	-	200
	Add for Gates, Signals, etc (Safety)	Job	670	900
	Add for Fence Enclosure (Safety)	Job	420	850
	Add for Platform Enclosure (Safety)	Job	320	500
	Hoist Engine, Gas Powered - 2 Drum	Month	-	1,700
	Add for Electric Powered	Month	-	200
	Towers and Hoists - Personnel 6000# Capacity			
	Based on 100' Unit - 8 1/2' Override	Month	-	3,400
	Add Above 100'	LnFt	-	32
	Installation - Up	LnFt	-	40
	Down	LnFt	-	36
.92	Cranes - Climbing - 20' Override			
	Based on 125' Bldg-130' Boom, 4000#	Month	-	5,800
	to 200' Boom, 7000#	Month	-	7,300
	230' Boom, 7000#	Month	-	9,800
	Add for LnFt Tower above 125'	Month	-	35
	Add for Free-Standing Type (Outside)	Month	-	750
	Installation			
	Foundation	Each	-	10,500
	440-Volt Transformer Install	Each	-	2,300
	Installation - Up (Outside-Inside)	LnFt	-	52
	Down (Outside)	LnFt	-	50
	Down (Inside)	LnFt	-	58
	Avg - Up, Down & Haul - 125' Tower	Each	-	29,000
	Add per Lift - Jacking	LnFt	-	60
	Add per Floor Pattern	-	470	400

.93	Cranes - Mobile Lifting (no Operator)	HOUR	DAY	MONTH
	15-Ton Hydraulic	95	850	3,400
	20-Ton	100	940	3,500
	25-Ton	105	1,020	5,400
	30-Ton	120	1,060	5,800
	45-Ton	135	1,200	6,600
	75-Ton Cable	155	1,300	7,800
	90-Ton	180	1,460	8,700
	25-Ton	210	1,800	12,500
	Add for Local Travel & Setup Time - 4 Hours			

*Subcontract

1-4

0103.0   MISCELLANEOUS CONDITIONS, Cont'd...

		UNIT	LABOR	MATERIAL
.94	Fork Lifts - Construction			
	Two-Wheel Drive - Gasoline Powered			
	21' 2500# Capacity	Month		1,800
	36' 4500# Capacity	Month		2,300
	Four-Wheel Drive - 21' 5000# Capacity	Month		2,300
	36' 6000# Capacity	Month		2,500
	38' 7000# Capacity	Month		2,800
	42' 8000# Capacity	Month		3,600
.95	Platform Lifts - Telescoping Boom - Rolling	Month		3,100
	Scissor - Rolling	Month		1,200
.10	TRUCKS AND OTHER JOB VEHICLES (Site Use)			
	$ 500,000 Job	JobAvg		1,500
	$1,000,000 Job	JobAvg		5,800
	$5,000,000 Job	JobAvg		13,000
	1-Ton Pickup	Month		500
	2-Ton Flatbed	Month		780
	2 1/2-Ton Dump	Month		1,300
.11	MAJOR EQUIPMENT (See Div. 2,3,4,5,& 6 if Subbed)			
	(See 1-6A & 7A: Rental Rates, Purchase Prices)			
	$ 100,000 Job	JobAvg		2,900
	$ 500,000 Job	JobAvg		7,100
	$1,000,000 Job	JobAvg		13,200
	$5,000,000 Job	JobAvg		52,000
	As a % of Labor (No Hoisting Equipment)			5%
.12	SMALL TOOLS AND SUPPLIES			
	(See 1-6A & 7A: Rental Rates, Purchase Prices)			
	$ 100,000 Job	JobAvg		1,200
	$ 500,000 Job	JobAvg		3,400
	$1,000,000 Job	JobAvg		5,700
	$5,000,000 Job	JobAvg		28,000
	As a Percentage of Labor	JobAvg		2%
.13	HAULING EQUIPMENT AND TOOLS			
	$ 100,000 Job	JobAvg	750	500
	$ 500,000 Job	JobAvg	1,500	1,350
	$1,000,000 Job	JobAvg	2,900	2,300
	$5,000,000 Job	JobAvg	9,500	7,000
.14	GAS AND OIL FOR EQUIPMENT			
	$ 500,000 Job	JobAvg		850
	$1,000,000 Job	JobAvg		1,300
	$5,000,000 Job	JobAvg		5,600
	As a Average of Equipment Cost	JobAvg		10%
.15	REPAIRS TO EQUIPMENT AND TOOLS			
	$ 100,000 Job	JobAvg		550
	$ 500,000 Job	JobAvg		1,100
	$1,000,000 Job	JobAvg		1,600
	$5,000,000 Job	JobAvg		6,700
	As an Average of Equipment Cost			10%
.16	PHOTOGRAPHS - 2 Views, 4 Copies	Month		110
.17	SCHEDULING - C.P.M., Pert, Network, Analysis			
	$ 500,000 Job	JobAvg		4,400
	$1,000,000 Job	JobAvg		7,500
	$5,000,000 Job	JobAvg		13,000
	Add for Update, Monitor	Each		680
.18	COMPUTER COSTS	Employee/Wk		8
.19	PERFORMANCE BOND - SUBCONTRACTORS	ContrAmt		1.5%
.20	PUNCH LIST AND FINAL OUT	ContrAmt		.5%

EQUIPMENT

		Approximate Cost New	Fair Rental* Rate per Mo.
	See 103.9 for Towers, Cranes and Lift Equipment		
Div. 1	**General Conditions**		
	Engineering Instruments - Transits	1,170	320
	Levels 18"	950	240
	Fans - 15,000 CFM - 36"	1,100	400
	Generators - 3500 Watt - Portable	1,200	420
	5000 Watt - Portable	1,550	675
	Heaters, Temp - Oil 100M to 300 BTU	1,300	525
	LP Gas 90M BTU	360	200
	150M BTU	400	270
	350M BTU	710	460
	Laser - Rotating	5,700	840
	Light Stands - Portable - 500 watt Quartz	670	250
	Pumps- Submersible 1/2 HP Elect 2"	420	160
	Centrifugal 2" 8,000 Gal	970	580
	3" 17,000 Gal	1000	590
	Diaphragm 2" - 3" x 3" - Pneumatic	1,400	440
	3" - 3" x 3" - Gas	1,450	570
	Submersible - 5 HP 3"	1,500	760
	Hose - Discharge - 3" - 25'	100	105
	Suction - 3" - 20'	130	135
	Vacuum Cleaners - Wet Dry - 5 Gal	600	160
Div. 2	**Excavation (Including Demolition)**		
	Air Tools- Breakers Light Duty - 35 to 40#	960	270
	Heavy Duty - 60 to 90#	1,320	375
	Rock Drill - 40#	1,780	390
	Chipping Hammer - 7# to 12#	770	340
	Clay Digger	720	300
	Hose - 3/4" - 50'	130	65
	Compressors 125 cfm Gas - Tow	10,700	770
	185 cfm Diesel - Trailer Mounted	13,600	1,080
	26' Conveyor - 2 HP Electric	9,150	1,175
	Electric Breakers - 60#	2,100	450
	Jacks - Hydraulic - 20 Ton - 11 HP	800	130
	Screw - 20 Ton	360	60
	Saws - Chain 18"	550	240
	Concrete Asphalt 14" Blade	1,600	400
	24" Plate - Vibratory Gas 170#	2,100	730
	Tampers - Rammer Gas 135#	2,300	780
Div. 3	**Concrete**		
	Bucket & Boot- 1 CuYd	2,050	400
	1 1/2 CuYd	2,600	420
	2 CuYd	3,400	440
	Buggies - Manual - 6 to 8 CuFt	470	115
	Power Operated - 16 CuFt	6,000	875
	Conveyors - 32' Hydraulic and Belt - Gas	9,400	1,200
	50' Hydraulic and Belt - Gas	20,800	2,500
	Core Drill - Diamond - 15 Amp	3,250	620
	with Trailer	9,400	1,025
	Crack Chaser	4,200	1,430
	Drill - Cordless 3/8" Reversing	290	95
	Heavy Duty 1/2" Reversing	370	200
	3/4" Reversing	670	300
	Sander- Hand - 9"	330	200
	Grinder- Ceiling 7' to 11' - Elec	2,200	260
	Floor - Elec - 1 1/2 HP	3,050	590
	Hammer - Drill - 3/4" Variable Speed	470	225
	Roto Chipping - 1 1/2"	500	420
	Hopper - Floor 30 CuFt	1,100	210
	Mixer - 6 CuFt	2,300	510
	Planer - 10'	4,300	720
	Scarifier & Scabblers - Single Bit	1,100	390
	5 Bit Floor - 175 cfm	7,800	2,900
	Stripper - Floor	3,750	560

*Weekly Rent 40% of monthly, Daily Rent 15% of monthly, Hourly Rent 2% of monthly.

EQUIPMENT

		Approximate Cost New	Fair Rental* Rate per Mo.
Div. 3	**Concrete, Cont'd...**		
	Saws - Rotary Hand - 7 1/4"	225	85
	9 1/4" - 8 1/4"	245	125
	Table 14" Radial Arm 3 HP	1,950	325
	Floor - Gas 11 HP	1,680	515
	Gas 18 HP	4,700	730
	Electric 8 HP 14"	2,100	580
	Cut off 14" Electric	760	500
	Screeds - Vibrating 6'	960	240
	12 1/2' Beam Type	2,075	375
	25' Beam Type	4,800	1,025
	Troweling Machines - Gas 36" 4 Blade 7 HP	1,750	520
	48" 4 Blade 6 HP	2,200	560
	Trunks & Tremmies - 12" and Hopper	210	70
	Vibrators - Flexible Head - 1 HP	1,200	390
	2 HP	1,250	460
	High cycle - 180 Cycle	3,800	540
	Back Panel 2 1/2 HP	820	240
	Wheelbarrows	125	50
Div. 4	**Masonry**		
	Hoist Electric - One ton	1,100	185
	Mixer - Mortar - 6 Cu Ft - Gas 7 HP	2,050	450
	Elec 1 1/2 HP	2,270	430
	8 Cu Ft - Gas 7 HP	2,500	575
	12 Cu Ft - Gas or Electric	3,800	840
	Pallet Jack	750	145
	Pump Grout - Hydraulic 10 GPM	6,900	1,550
	Saws - Table - 1 1/2 HP - 14"	2,300	325
	5 HP - 14"	3,450	620
	Scaffold - 5' - 6" with Braces	80	3.90
	6' - 6" with Braces	85	4.40
	Wheels	90	12.00
	Brackets	25	2.00
	Plank - Aluminum	100	6.00
	16' Laminated	40	3.20
	Base - Adj	25	1.75
	Splitter - 10,000#	1,750	380
	Rolling Towers	380	100
	Swing Stage - Motorized w/ deck-cables-safety	10,000	1,200
Div. 6	**Carpentry**		
	Door Hanging Eq. (Router, Mortiser, Template)	300	100
	Drills - Heavy Duty (See Div. 3)		
	Fasteners - Power - Stud Gun	275	100
	Hammer - Drill 1/2"	315	125
	Miter Box - 14" Power	430	150
	Nailers and Staplers - Spot w/ compressor	1,850	270
	Planers	600	155
	Sanders - Disc 9"	310	100
	Belt 3" - Heavy Duty	420	125
	Floor	450	90
	Saws - Band	300	125
	Chain	420	250
	Rotary, Hand - 7 1/4"	210	110
	8 1/4"	240	125
	Table - Fixed 10"	420	150
	Jig	185	75
	Reciprocating	450	240
	Cut-off - 12" Elec.	2,200	490
	Scaffold - Rolling	500	115
	Screwdriver - Variable Speed - Drywall	360	125
	Stud Drivers - Low Velocity	320	140
	High Velocity	460	155
	Wrench - 1/2" Impact - Air	330	155
	1" Impact - Air	680	270

*Weekly Rent 40% of monthly, Daily Rent 15% of monthly, Hourly Rent 2% of monthly.

		PAGE
0201.0	**DEMOLITION & CLEARING (CSI 02100)**	**2-3**
.1	STRUCTURE MOVING	2-3
.2	CLEARING & GRUBBING	2-3
.3	DEMOLITION	2-3
.31	Total Building	2-3
.32	Selective Building	2-3
.33	Site	2-4
.34	Disposal Haul Away	2-4
.35	Cutting, Core Driller and Blasting	2-4
0202.0	**EARTHWORK (CSI 02200)**	**2-5**
.1	GRADING	2-5
.2	EXCAVATION	2-5
.3	BACKFILL	2-5
.4	COMPACTION	2-5
.5	DISPOSAL	2-6
.6	TESTS AND SUBSURFACE EXPLORATION	2-6
0203.0	**PILE AND CAISSON FOUNDATIONS (CSI 02300)**	**2-6**
.1	CONCRETE PILING	2-6
.11	Augered and Cast-In-Place Concrete	2-6
.12	Precast Prestressed	2-6
.2	STEEL	2-6
.21	Steel Pipe	2-6
.22	Steel H. Section	2-6
.3	WOOD	2-6
.4	CAISSONS	2-7
0204.0	**EARTH AND WATER RETAINER WORK (CSI 02400)**	**2-7**
.1	STREET SHEET PILING	2-7
.2	WOOD SHEET PILING AND BRACING	2-7
.3	H. PILES AND WOOD SHEATHING	2-7
.4	UNDERPINNING	2-7
.5	RIP RAP AND LOOSE STONEWALLS	2-7
.6	CRIBBING	2-7
0205.0	**SITE DRAINAGE (CSI 02500)**	**2-7**
	(See Divisions 15 and 16 for Site Utilities)	
.1	DITCHING AND BACKFILL (OPEN)	2-7
.2	DEWATERING	2-7
.3	CULVERTS	2-8
.4	CATCH BASINS AND MANHOLES	2-8
.5	FOUNDATION DRAINAGE	2-8
0206.0	**PAVEMENTS, CURBS AND WALKS (CSI 02600)**	**2-9**
.1	PAVING	2-9
.11	Bituminous	2-9
.12	Concrete	2-9
.13	Stabilized Aggregate	2-9
.14	Brick	2-9
.2	CURBS AND GUTTERS	2-10
.21	Concrete - Cast in Place	2-10
.22	Concrete - Precast	2-10
.23	Bituminous	2-10
.24	Granite	2-10
.25	Timbers, Treated	2-10
.26	Plastic	2-10

0206.0   **PAVEMENTS AND WALKS, CONT'D...**

    .3   WALKS   2-10
    .31   Bituminous   2-10
    .32   Concrete   2-10
    .33   Precast Block   2-10
    .34   Crushed Rock   2-10
    .35   Flagstone   2-10
    .36   Brick   2-10
    .37   Wood   2-10

0207.0   **FENCING**   **2-11**

    .1   CHAIN LINK   2-11
    .2   WELDED WIRE   2-11
    .3   MESH WIRE   2-11
    .4   WOOD   2-11
    .5   GUARD RAILS   2-11

0208.0   **RECREATIONAL FACILITIES**   **2-12**

    .1   PLAYING FIELDS AND COURTS   2-12
    .2   RECREATIONAL EQUIPMENT   2-12
    .3   SITE FURNISHINGS   2-12
    .4   SHELTERS   2-12

0209.0   **LANDSCAPING (CSI 02900)**   **2-13**

    .1   SEEDING   2-13
    .2   SODDING   2-13
    .3   DISCING   2-13
    .4   FERTILIZING   2-13
    .5   TOP SOIL   2-13
    .6   TREES AND SHRUBS   2-13
    .7   BEDS   2-13
    .8   WOOD CURBS AND WALLS   2-13

0210.0   **SOIL STABILIZATION**   **2-14**

    .1   CHEMICAL INTRUSIONS   2-14
    .2   CONCRETE INTRUSIONS   2-14
    .3   VIBRO-FLOTATION   2-14

0211.0   **SOIL TREATMENT**   **2-14**

    .1   PESTICIDES   2-14
    .2   VEGETATION   2-14

0212.0   **RAILROAD WORK (CSI 02450)**   **2-14**

    .1   NEW WORK   2-14
    .2   REPAIR WORK   2-14

0213.0   **MARINE WORK (CSI 02480)**   **2-14**

0214.0   **TUNNEL WORK (CSI 02300)**   **2-14**

     See Division 1501 for Piped Utility Material   (CSI 02600)
     See Division 1501 for Water Distribution   (CSI 02660)
     See Division 1501 for Fuel Distribution   (CSI 02680)
     See Division 1503 for Sewage Distribution   (CSI 02700)
     See Division 1600 for Power Communication   (CSI 02780)

0201.0	**DEMOLITION & CLEARING (Op.Eng., Truck Dr. & Lab.)**	UNIT	COST with MACHINE
.1	STRUCTURE MOVING - Wood Frame	SqFt	13.00
	With Masonry	SqFt	16.00
.2	CLEARING AND GRUBBING		
	Trees - Trunk	Dia/Inch	13.00
	Stump - Removed	Dia/Inch	10.00
	Chipped - Below Grade	Dia/Inch	5.90
	Add for Hauling Away and Dumping	Dia/Inch	4.60
	Shrubs & Brush - Light	Acre	1,750.00
	Medium	Acre	2,500.00
	Heavy	Acre	3,300.00
.3	DEMOLITION AND LOAD (See .34 for Disposal)		
.31	Total Building		
	Wood - Mainly Housing	CuFt/Bldg	.27
	Masonry - Load Bearing	CuFt/Bldg	.33
	Concrete - Frame	CuFt/Bldg	.41
	Steel - Frame	CuFt/Bldg	.33
	Add for Floor Above 2 Stories	CuFt/Bldg	.06

.32 Selective Building Removals (No Cutting/Disposal)
(See .35 - Cut and Drill)
(See .34 - Disposal/Haul Away)
(See .33 - Site Removal)

	Unit	Labor Only	Machine Only	Unit	Labor Only	Machine Only
Concrete:						
8" Walls Reinforced	CuYd	230.00	88.00	SqFt	5.75	1.65
Non-Reinforced	CuYd	180.00	77.00	SqFt	4.40	1.43
12" Walls Reinforced	CuYd	310.00	67.00	SqFt	11.20	2.50
Footings - 24" x 12"	CuYd	250.00	150.00	LnFt	9.20	5.50
6" Struc. Slab Reinforced	CuYd	195.00	60.00	SqFt	3.60	1.09
8" Struc. Slab Reinforced	CuYd	208.00	80.00	SqFt	5.15	1.48
4" Slab on Ground Reinforced	CuYd	370.00	50.00	SqFt	1.65	.59
Non-Reinforced	CuYd	330.00	44.00	SqFt	1.28	.55
6" Slab on Ground Reinforced	CuYd	210.00	41.00	SqFt	2.00	.74
Non-Reinforced.	CuYd	200.00	38.00	SqFt	1.50	.70
9" Stairs Reinforced	CuYd	268.00	88.00	SqFt	6.20	2.40
Masonry:						
4" Brick or Stone Walls	CuYd	96.00	30.00	SqFt	1.18	.36
and 8" Backup	CuYd	175.00	66.00	SqFt	2.15	.66
Block or Tile Partitions	CuYd	75.00	29.00	SqFt	.98	.33
6" Block or Tile Partitions	CuYd	60.00	22.00	SqFt	1.08	.40
8" Block or Tile Partitions	CuYd	52.00	41.00	SqFt	1.29	.50
12" Block or Tile Partitions	CuYd	140.00	55.00	SqFt	1.65	.60
Add for Plastered Type				SqFt	.36	.11

Misc - Hand Work (Add for Equipment & Scaffold)	UNIT	Labor	
Acoustical Ceilings - Attached (Incl. Iron)	SqFt	.52	-
Suspended (Incl. Grid)	SqFt	.33	-
Asbestos - Ceilings and Walls	SqFt	13.40	-
Columns and Beams	SqFt	33.00	-
Pipe	LnFt	41.50	-
Tile Flooring	SqFt	1.55	-
Cabinets and Tops	LnFt	9.30	-
Carpet	SqFt	.26	-
Ceramic and Quarry Tile	SqFt	1.05	-
Doors and Frames - Metal	Each	46.50	-
Wood	Each	41.50	-
Drywall Ceilings - Attached (Including Grid)	SqFt	.67	-
Wood or Metal Studs - 2 Sides	SqFt	.72	-
Paint Removal - Doors and Windows	SqFt	.67	-
Walls	SqFt	.52	-
Plaster on Wood or Metal Studs	SqFt	.98	-
Ceilings - Attached (Incl. Iron)	SqFt	.93	-
Roofing - Built-Up	SqFt	.88	-
Shingles - Asphalt and Wood	SqFt	.29	-
Terrazzo Flooring	SqFt	1.55	-
Vinyl Composition Flooring (No Mastic Removal)	SqFt	.31	-
Wall Coverings	SqFt	.57	-
Windows - Metal or Wood	Each	31.00	-
Wood Flooring	SqFt	.33	-

0201.0 DEMOLITION & CLEARING, Cont'd...

.33   Site Removals

(See .34 for Disposal)

(See .35 for Cutting Additions)

	UNIT	LABOR ONLY (1)	MACHINE ONLY
Concrete: 4" Sidewalks - Reinforced	SqFt	1.75	.60
Non -Reinforced	SqFt	1.40	.57
6" Drives - Reinforced	SqFt	2.10	.77
Non-Reinforced	SqFt	1.64	.82
Curb - 6" x 18"	LnFt	4.85	1.45
and Gutter	LnFt	7.80	1.95
Asphalt: 2"	SqFt	1.05	.36
Curb	LnFt	2.50	.65
Fencing: 8" Metal	LnFt	1.60	.40

.34   Disposal or Haul Away (Including Truck or Box & Driver)

Haul - No Dump Charges Included

	UNIT	COST
Concrete Truck Measure - 75¢ a minute or	CuYd/Mile	.70
Masonry - 75¢ a minute or	CuYd/Mile	.66
Wood - 75¢ a minute or	CuYd/Mile	.60
Box Rent (Included Above) - 30 Cu.Yd.	Box	360.00
Clean Material - 20 Cu.Yd.	Box	320.00
10 Cu.Yd.	Box	270.00
Dump Charges - Concrete	CuYd	6.50
Building Materials	CuYd	9.50
Tree and Brush Removal	CuYd	13.00

.35   Saw Cutting/Core Drilling/Torch Cutting/Blasting

Saw Cutting (includes Operator):

	UNIT		COST
Slabs - Asphalt	LnFt	per inch	1.10
Concrete - Structural	LnFt	per inch	2.00
Hollow	LnFt	per inch	.85
Slabs on Grd/Non - Rein.	LnFt	per inch	.75
/with Mesh	LnFt	per inch	.95
Walls - Concrete - Reinforced	LnFt	per inch	3.10
Non - Reinforced	LnFt	per inch	2.50
Masonry - Hollow	LnFt	per inch	1.80
Solid	LnFt	per inch	2.30

Core Drilling:

	UNIT	COST
Solid Slab - 2" Diameter	Inch	3.55
3" Diameter	Inch	3.85
4" Diameter	Inch	5.40
6" Diameter	Inch	7.80
8" Diameter	Inch	10.50
Hollow Core Slab - 2" Diameter	Inch	2.30
3" Diameter	Inch	2.50
4" Diameter	Inch	3.65
6" Diameter	Inch	6.00
Torch Cutting - Steel per Inch	LnFt	5.00
Blasting - Unrestricted	CuYd	105.00
Restricted	CuYd	190.00

Cut Openings:

	UNIT	COST
Concrete (see .32)	SqFt	-
Sheet Rock Partitions	SqFt	2.90
Wood Floors	SqFt	2.30

No Disposal or Haul Away Included (see .34)

---

(1) With Pneumatic Tools

0202.0 EARTHWORK (Op. Eng., Truck Driver & Lab.)

.1 GRADING

Machine	UNIT(1)	COST	UNIT(1)	COST
Strip Top Soil - 4"	CuYd	5.00	SqYd	.55
Spread Top Soil				
4" Site Borrow	CuYd	5.45	SqYd	.60
Off-Site Borrow ($12 CuYd)	CuYd	19.50	SqYd	2.20

Rough Grading - Cut and Fill	UNIT	SOFT(1)	MED.(1)	HARD(1)
Dozer - To 200'	CuYd	3.30	3.55	4.20
To 500'	CuYd	3.70	5.30	6.60
Scraper - Self Propelled - To 500'	CuYd	2.95	3.30	3.85
To 1,000'	CuYd	3.75	4.40	5.20
Grader - To 200'	CuYd	3.25	3.80	5.50
Add for Compaction (See 0202.4)				
Add for Hauling Away (See 0202.5)				
Add for Truck Haul on Site	CuYd	2.55	-	-
Add for Borrow Brought In				
Loose Fill - 2.75 Ton	CuYd	10.40	-	-
Crushed Stone - 5.30 Ton	CuYd	13.75	-	-
Gravel - 4.10 Ton	CuYd	12.00	-	-

Hand and Machine	UNIT	HAND LABOR	MACHINE
Fine Grading - 4" Fill - Site	CuYd	10.80	5.20
or SqFt		.14	.07
4" Fill - Building	CuYd	20.00	9.80
or SqFt		.25	.12

.2 EXCAVATION - Dig & Cast or Load (No Hauling)

Soil

	UNIT	LABOR	MACHINE
Open - Soft (Sand) w/Backhoe or Shovel	CuYd	19.60	3.20
Medium (Clay)	CuYd	27.75	3.95
Hard	CuYd	43.20	5.50
Add for Clamshell or Dragline	CuYd	-	1.25
Deduct for Front End Loader	CuYd	-	1.05
Deduct for Dozer	CuYd	-	.27
Trench or Pocket - Soft w/Backhoe or Shovel	CuYd	22.60	4.70
Medium	CuYd	28.70	6.10
Hard	CuYd	41.30	7.40
Add for Clamshell or Dragline	CuYd	-	1.15
Augered - 12" diameter	CuYd	47.40	9.00
24" diameter	CuYd	36.20	6.10
Add for Frost	CuYd	-	7.20
Rock - Soft - Dozer	CuYd	-	31.00
Hammer	CuYd	-	77.00
Medium - Blast	CuYd	-	93.00
Hammer	CuYd	-	160.00
Hard - Blast	CuYd	-	205.00
Hammer	CuYd	-	275.00
Swamp	CuYd	-	8.90
Underwater	CuYd	-	11.50

.3 BACKFILL (Not Compacted and Site Borrow)
	UNIT	LABOR	MACHINE
	CuYd	10.05	2.50
Add for Off-Site Borrow	CuYd	-	10.35

Add to All Above for Mobilization Average   5%

(1) Machine and Operators

0202.0 EARTHWORK, Cont'd...

		UNIT	LABOR (1)	COST (2)
.4	COMPACTION (Incl. Backfilling & Site Borrow)			
	Labor Only			
	Building to 85 - 24" Lifts	CuYd	14.50	-
	to 95 - 12" Lifts	CuYd	17.00	-
	to 100 - 6" Lifts	CuYd	20.30	-
	Labor Compaction & Machine Backfill Combined			
	Building to 85 - 24" Lifts	CuYd	8.85	2.60
	to 90 - 18" Lifts	CuYd	9.40	2.70
	to 95 - 12" Lifts	CuYd	9.90	2.80
	to 100 - 6" Lifts	CuYd	10.80	3.25
	Subgrade (Including Grading and Rollers)	CuYd	-	3.65
	Machine Only (No Grading)			
	Site to 85 - 24" Lifts	CuYd		3.10
	to 90 - 18" Lifts	CuYd		3.20
	to 95 - 12" Lifts	CuYd		3.70
	to 100 - 6" Lifts	CuYd		4.35
	Add to Above for Off-site Borrow	CuYd		10.20
.5	DISPOSAL (Haul Away) Country	CuYd/Mile		.65
	City	CuYd/Mile		.76
.6	TESTS			
	Field Density - Compaction	Each		35.00
	Soil Gradation	Each		51.00
	Density (Proctor)	Each		82.00
	Moisture and Specific Gravities	Each		87.00
	Hydrometer (Sand, Silt and Clay)	Each		100.00
	Rock -2" Core	LnFt		63.00
	4" Core	LnFt		105.00
	Add for Soil Penetration	LnFt		7.15
	Add for Mileage	Mile		.41

0203.0 PILE AND CAISSON FOUNDATIONS (L&M)
(Operating Engineers and Iron Workers)

		UNIT	COST
.1	CONCRETE		
.11	Augured or Cast in Place -12"	LnFt	20.00
.12	Precast Prestressed -10"	LnFt	17.70
	12"	LnFt	20.50
	14"	LnFt	22.50
	16"	LnFt	27.00
.13	Pipe Casing Pulled - Cast - in - Place - 10"	LnFt	18.30
	12"	LnFt	19.40
.2	STEEL		
.21	Steel Pipe -10"	LnFt	23.00
	12"	LnFt	27.00
	14"	LnFt	31.00
	16"	LnFt	34.00
	Concrete - Filled - 10"	LnFt	26.50
	12"	LnFt	32.00
	14"	LnFt	38.00
	16"	LnFt	43.00
.22	Steel H Section 8" - 36#	LnFt	24.00
	10" - 42#	LnFt	27.00
	12" - 53#	LnFt	33.00
	14" - 73#	LnFt	35.00
	14" - 89#	LnFt	42.00
.3	WOOD (Treated) to 40' (Incl. Tests & Cutoffs)		
	10"	LnFt	17.00
	12"	LnFt	18.00
	14"	LnFt	20.00
	16"	LnFt	21.00
	Add for Mobilization		10%

---

(1) Hand Work
(2) Machine and Operators

		UNIT	COST
0203.0	**PILE & CAISSON, Cont'd...**		
.4	CAISSONS -20" Concrete Filled	LnFt	34.00
	24"	LnFt	41.00
	30"	LnFt	55.00
	36"	LnFt	70.00
	42"	LnFt	91.00
	48"	LnFt	115.00
	54"	LnFt	150.00
	60"	LnFt	185.00
	Add for Bells	Each	1,750.00
	Add for Wet Ground	LnFt	5.00
	Add for Obstructions - Hand Removal	CuFt	57.00
	Add for Obstructions - Machine Removal	CuFt	10.00
	Add for Reinforced Steel	Ton	1,500.00
	Add for Mobilization		10%
0204.0	**EARTH & WATER RETAINER WORK**		
.1	STEEL SHAFT PILING (Including Bracing) (L&M)		
	Heavy -27# -12"-16' Full Salvage	SqFt	16.00
	40# -15"-16' Full Salvage	SqFt	17.00
	57# -18"-16' Full Salvage	SqFt	18.00
	Add per Foot Longer than 16'	SqFt	2.00
	Add for No Salvage	SqFt	8.00
	Light -12 ga - 18" Black to 8' Full Salvage	SqFt	14.00
	10 ga - 18" Black to 8' Full Salvage	SqFt	15.00
	8 ga - 18" Black to 8' Full Salvage	SqFt	19.00
	Add for No Salvage	SqFt	7.00
	Add for Galvanized	SqFt	30%
.2	WOOD SHEET PILING & BRACING		
	6' Deep	SqFt	5.00
	8' Deep	SqFt	5.25
	10' Deep	SqFt	5.70
	12' Deep	SqFt	6.25
.3	H PILES & WOOD SHEATHING (4" Lbr.)		
	10' Deep	SqFt	19.00
	15' Deep	SqFt	20.00
	20' Deep	SqFt	25.00
	25' Deep	SqFt	29.00
	Deduct for Salvage	SqFt	40%
.4	UNDERPINNING	CuYd	650.00
.5	RIP RAP & LOOSE STONE WALLS		
	Machine Placed	CuYd	15.50
	Hand Placed	CuYd	50.00
	and Grouted	CuYd	120.00
.6	CRIBBING		
	6" Open to 10'	SqFt	22.00
	8" Open	SqFt	26.00
0205.0	**SITE DRAINAGE (Including Building Foundations)**		
.1	DITCHING AND BACKFILL (Open) - See 0202.2		
.2	DEWATERING		
	Pumping -2" Pump	Day	53.00
	3" Pump	Day	82.00
	4" Pump	Day	110.00
	Well Points - Ln.Ft. Header - first month	Month	125.00
	Add for Each Additional Month	Month	65.00
	Add for Attending Time	Hour	44.00

0205.0   SITE DRAINAGE, Cont'd...

.3   CULVERTS & PIPING (No Excavation - See 0202.2)

.31   Corrugated Galvanized Metal

	UNIT	LABOR	MATERIAL
Watertight (Bituminous Coated)			
12" Diameter w/Bands & Gaskets	LnFt	3.85	14.00
15"	LnFt	4.15	18.00
18"	LnFt	4.68	20.00
24"	LnFt	5.50	25.00
30"	LnFt	6.55	32.00
36"	LnFt	7.75	39.00
48"	LnFt	16.50	45.00
Non - Watertight			
12" Diameter w/Bands & Gaskets	LnFt	3.10	7.00
15"	LnFt	3.70	8.00
18"	LnFt	4.50	10.00
24"	LnFt	6.20	12.00
30"	LnFt	7.60	15.00
36"	LnFt	8.80	23.00
48"	LnFt	7.85	25.00
Add for Oval Arch Shape	LnFt	20%	-
Aprons			
18"	Each	15.50	65.00
24"	Each	17.50	95.00
30"	Each	19.50	170.00
36"	Each	22.70	270.00

.32   Reinforced Concrete Pipe - #3 - Class 5

	UNIT	LABOR	MATERIAL
12" Diameter	LnFt	5.15	12.50
18"	LnFt	6.90	19.60
24"	LnFt	9.70	27.00
27"	LnFt	11.80	39.20
30"	LnFt	15.00	48.50
36"	LnFt	18.20	57.00
48"	LnFt	26.80	88.00
Add for Each Added Foot in Depth	LnFt	5%	-
Add for Apron 36" (Flared End)	Each	155.00	520.00
Add for Trash Guards (Galvanized)	Each	135.00	830.00

.33   Vitrified Clay 12"

	UNIT	LABOR	MATERIAL
Vitrified Clay 12"	LnFt	5.40	11.20
24"	LnFt	8.65	43.80

.4   CATCH BASINS & MANHOLES (No Excavation)

	UNIT	LABOR	MATERIAL
48" Cast in Place (Concrete)	LnFt	140.00	145.00
48" Brick or Block	LnFt	135.00	150.00
48" Precast Concrete	LnFt	88.00	86.00
48" Precast Concrete Collar	LnFt	115.00	150.00
Add for Cover - Precast Concrete	Each	62.00	135.00
Add for Base - Precast Concrete	Each	83.00	140.00
Add for Covers & Grates - H.D.C.I.	Each	36.00	160.00
Add for Covers & Grates - L.D.C.I.	Each	36.00	160.00
Add for Covers - H.D. Watertight	Each	36.00	275.00
Add for Steps	Each	-	14.00
Add for Adjusting Rings 2"	Each	13.30	14.00

.5   FOUNDATION DRAINAGE (Subdrainage)

	UNIT	LABOR	MATERIAL
4" Clay Pipe	LnFt	1.05	1.55
4" Corrugated - Perforated	LnFt	.88	.95
4" Plastic Pipe - Perforated	LnFt	.88	1.15
Solid	LnFt	.88	.90
6" Clay Pipe	LnFt	1.30	1.75
6" Corrugated - Perforated	LnFt	1.05	1.35
6" Plastic Pipe - Perforated	LnFt	1.05	2.25
Solid	LnFt	.98	1.80
Add for Porous Surround 2'x 2' ($8.00 CuYd)	LnFt	1.30	2.60

0206.0   PAVEMENT, CURBS AND WALKS

   .1   PAVING (PARKING LOTS AND DRIVEWAYS)

		UNIT	COST
.11	Bituminous (Material $35/Ton, 5% Asphalt, 10-Mile Delivery)		
	1 1/2" Wearing 10.5 SqYd/Ton	SqYd	4.50
	2" Wearing 9.0 SqYd/Ton	SqYd	5.70
	2 1/2" Wearing 7.5 SqYd/Ton	SqYd	6.50
	3" Wearing 6.0 SqYd/Ton	SqYd	7.60
	Add for Paths and Small Driveways	SqYd	4.20
	Add for Patching	SqYd	9.50
	Add for Asphalt Content per 1%	SqYd	2.00
	Add for Over 10-mile Delivery - Ton/Mile	SqYd	.35
	Deduct for Base Asphalt	SqYd	10%
	Add for Seal Coat	SqYd	1.50
	Add for Striping - Paint	LnFt	.25
	Add for Striping - Plastic	LnFt	.50
	Base Course for Above - Add:		
	4" Sand & Gravel $8.80 Ton and 1/3 Ton	SqYd	4.00
	4" Crushed Stone $10.90 Ton	SqYd	4.80
	6" Sand & Gravel $7.70 Ton and 1/2 Ton	SqYd	4.90
	6" Crushed Stone $10.00 Ton	SqYd	5.80
	Add per Mile over 10 Miles	SqYd	.40
	Deduct for Belly Dump Delivery - $.75 Ton	SqYd	.40
	4" Sand & Gravel	Ton	8.80
	4" Crushed Stone	Ton	10.00
	6" Sand & Gravel	Ton	8.20
	6" Crushed Stone	Ton	9.20
	Add per Ton Mile Over 10 Miles	Ton	.40
	Deduct for Belly Dump Delivery	Ton	.90
.12	Concrete (4000# Ready Mix, $74 CuYd - Machine Placed)		
	6" Reinforced with 6 x 6, 8-8 Mesh	SqYd	21.00
	7" Reinforced with 6 x 6, 6-6 Mesh	SqYd	22.00
	8" Reinforced with #4 Rods - 12" O.C.	SqYd	25.70
	9" Reinforced with #5 Rods - 12" O.C.	SqYd	26.70
	10" Reinforced with #6 Rods - 8" O.C.	SqYd	28.30
	12" Reinforced with #8 Rods - 8" O.C.	SqYd	36.00
	Add for Base - Same as Base Prices Above		
	Add to Above for Hand Placed	SqYd	25%
	Add for Driveways	SqYd	50%
.13	Stabilized Aggregate (Delivery by Dump Truck)		
	4" Gravel - $8.80 Ton - 3/4" Compacted	SqYd	4.80
	4" Crushed Stone - $10.00 Ton - 3/4" Compacted	SqYd	5.80
	6" Gravel - $8.00 Ton - 3/4" Compacted	SqYd	5.60
	6" Crushed Stone - $9.30 Ton - 3/4" Compacted	SqYd	7.30
	6" Pulverized Concrete	SqYd	7.80
	Add for Soil Cement Treatment	SqYd	.93
	Add for Oil Penetration Treatment	SqYd	2.80
	Add for Calcium Chloride Treatment	SqYd	.48
	Add per CuYd Mile Over 10 Miles	SqYd	.40
	Deduct for Belly Dump Truck Delivery	SqYd	.40
.14	Brick - 4" x 8" x 2 1/2"	SqYd	75.00
	Add for Sand Bed - Compacted	SqYd	6.40
	Add for Mortar Setting - Bed & Joints	SqYd	7.00

0206.0 PAVEMENT, CURBS AND WALKS, Cont'd...

.2 CURBS & GUTTERS (Cost Incl. Excavation & Equip.)

	UNIT	LABOR	MATERIAL
.21 Concrete - Cast in Place			
Curb			
6" x 12" Small Job - 200' Day	LnFt	5.25	2.70
6" x 18"	LnFt	6.40	3.20
6" x 24"	LnFt	8.00	4.20
6" x 30"	LnFt	9.25	5.00
Add for Reinforced - Two #5 Rods	LnFt	.56	.88
Add for Gutter - 6" x 12"	LnFt	3.75	2.10
Add for Gutter - 6" x 18"	LnFt	4.15	2.95
Add for Large Job - Over 200' Day	LnFt		10%
Add for Formless 1000' Day Up	LnFt		20%
Add for Curved or Radius Work	LnFt		40%
Curb & Gutter Rolled			
6" x 12" Small Job	LnFt	6.60	4.05
6" x 18"	LnFt	7.75	5.50
6" x 24"	LnFt	9.40	7.00
Deduct for Large Job - Over 200' Day	LnFt	-	10%
Deduct for Formless 1000' Day Up	LnFt	-	25%
Add for Reinforced - Two #5 Rods	LnFt	.62	.75
Add for Curved or Radius Work	LnFt	-	40%
Add for Approaches - 8' x 16' x 6'	SqFt	3.35	2.00
.22 Concrete - Precast & Pinned - 6" x 6'	LnFt	1.75	6.80
6" x 7'	LnFt	1.95	6.30
6" x 8'	LnFt	1.85	5.75
.23 Asphalt - 6" x 8'	LnFt	.88	1.25
Add for Curved Work	LnFt	-	30%
.24 Granite - 6" x 16"	LnFt	4.95	22.00
.25 Timbers, Treated - 6" x 6"	LnFt	1.75	3.30
6" x 8"	LnFt	2.28	4.80
.26 Plastic - 6" x 6"	LnFt	1.55	4.75
.3 WALKS AND DRIVEWAYS			
Bituminous - 1 1/2" with 4" base	SqFt	.39	.57
2" with 4" base	SqFt	.45	.60
Concrete - 4" - Broom Finish	SqFt	1.40	1.10
5" - Broom Finish	SqFt	1.42	1.30
6" - Broom Finish	SqFt	1.47	1.35
Add Mesh 6" x 6", 10 - 10	SqFt	.10	.14
Add for Exposed Aggregate	SqFt	.67	.14
Add for Coloring (Acrylic)	SqFt	.35	.61
Add for Base - 4" Sand or Gravel	SqFt	.24	.20
Precast Block - Colored 1" with 2" Sand Cushion	SqFt	1.30	.80
Colored 2" with 2" Sand Cushion	SqFt	1.50	1.00
Crushed Rock - 4" Compacted	SqFt	.29	.27
Flagstone - 1 1/4" with Sand Bed	SqFt	3.50	6.90
1 1/4" with Mortar Setting Bed	SqFt	4.85	7.90
Brick - 4" x 8" x 2 1/4" with 4" Sand Bed	SqFt	3.40	2.75
4" x 8" x 2 1/4" with Mortar Setting Bed	SqFt	3.95	3.50
Add for Herringbone or Weave Pattern	SqFt	1.25	1.30
Wood - 2" T&G on 6" x 6" Timbers	SqFt	1.40	2.40
2" Boards on 4" x 4" Timbers	SqFt	1.25	2.05
Asphalt Block - 6" x 12" x 3"	SqFt	2.20	4.00
Slate - 1 1/4"	SqFt	5.25	7.40
.4 STAIRS - EXTERIOR	SqFt	6.20	1.30

0207.0 FENCING (L&M) (Ironworkers or Carpenters)

	UNIT	COST
.1 CHAIN LINK (Galvanized 9 ga) Including Intermediate		
Posts and Concrete Embedded		
5' High - 2" Mesh and 2" O.D. Pipe	LnFt	12.00
6' High	LnFt	13.00
7' High	LnFt	13.50
8' High	LnFt	16.20
10' High	LnFt	30.00
Add for 1 - or 3 - Strand Barbed Wire	LnFt	2.60
Add for Vinyl Coated Wire	LnFt	15%
Add for Aluminum Wire	LnFt	25%
Add for Wood Slats	LnFt	50%
Add for Aluminum Slats	LnFt	75%
Add for Posts - Corner, End and Gate:		
5' High - 3" O.D.	Each	39.00
6' High - 3" O.D.	Each	44.00
7' High - 3" O.D.	Each	55.00
8' High - 3" O.D.	Each	64.00
10' High - 3" O.D.	Each	90.00
Add for Gate Frames:		
5' High	LnFt	38.00
6' High	LnFt	40.00
7' High	LnFt	50.00
8' High	LnFt	64.00
10' High	LnFt	78.00
.2 WELDED WIRE (Galvanized 11 ga) Including Intermediate		
Posts and Concrete Embedded		
4' High - 2" x 4"	LnFt	9.80
5' High	LnFt	11.00
6' High	LnFt	12.20
Add for Posts:		
4' Long	Each	22.00
5' Long	Each	23.00
6' Long	Each	25.00
Add for Gate Frames:		
4' Long	LnFt	21.00
5' Long	LnFt	25.00
6' Long	LnFt	27.00
Add for Slats	LnFt	5.00
.3 MESH WIRE (Galvanized 14 ga)		
3' High	LnFt	5.00
4' High	LnFt	5.80
5' High	LnFt	6.60
.4 WOOD FENCING		
Rail - 4' Cedar - 3 Boards	LnFt	14.00
Redwood - 3 Boards	LnFt	14.50
Pine, Painted - 3 Boards	LnFt	13.50
Vertical - 6' Pine, Painted	LnFt	17.00
Redwood	LnFt	21.00
Cedar	LnFt	22.00
Weave - 6' Redwood	LnFt	22.50
8' Redwood	LnFt	25.00
Picket - 4' Cedar	LnFt	10.80
6' Cedar	LnFt	14.00
Split Rail - 4' - 3 Rail	LnFt	7.50
Add for Post Set in Concrete	Each	5.00
.5 GUARD RAILS - Corrugated - Galvanized with Wood Posts	LnFt	27.00
Galvanized with Steel Posts	LnFt	29.00
Wire Cable - with Wood Posts, 3 - Strand	LnFt	14.00
with Steel Posts, 3 - Strand	LnFt	15.00
with no posts, 3-strand	LnFt	7.00

0208.0 RECREATIONAL FACILITIES

		UNIT	COST
.1	PLAYING FIELDS AND OUTDOOR COURTS		
.11	Football Fields - Field Size - 300' x 160		
	Minimum Use Size - 306' x 180'		
	Artificial Turf	Each	600,000.00
	Natural Turf	Each	275,000.00
.12	Tennis Courts -Court Size - 78' x 36'		
	Minimum Use Size - 108' x 48'		
	Asphalt with Color Surfacing - 1 Court	Each	22,000.00
	2 Courts	Each	18,000.00
	3 Courts	Each	17,500.00
	Incl. 3" Asphalt, 4" Gravel Base & Striping		
	Synthetic	Each	32,000.00
	Concrete	Each	26,500.00
	Add for Practice Boards	Each	2,000.00
	Add for Net Posts (4" Pipe with Ratchet)	Pair	550.00
	Add for Steel nets	Each	550.00
.13	Track and Field		
	Track - 440 yard - Track Size 21' x 1,320'		
	Synthetic Turf - 1" Polyurethane	Each	130,000.00
	1" Polyurethane	SqYd	36.00
	3" Bituminous Base	SqYd	7.00
	6" Gravel Base	SqYd	4.75
	Rubberized Asphalt	Each	61,000.00
	1" with 12% Resiliency	SqYd	10.50
	Add Bituminous Base - same as above	SqYd	6.00
	Cinder	Each	50,000.00
	High Jump - 50' Radius w/12' x 16' P.T.	Each	9,400.00
	Long Jump - 200' x 4' Runway w/20' x 8' P.T.	Each	4,150.00
	Pole Vault - 130' x 4' Runway w/16' x 16' P.T.	Each	2,600.00
.14	Basketball Courts -		
	Concrete - 4" Grade School - Court 42' x 74'	Each	15,500.00
	High School - Court 50' x 84'	Each	19,000.00
	College - Court 50' x 94'	Each	22,000.00
	Asphalt - 1 1/2" - Grade School	Each	7,500.00
	High School	Each	8,500.00
	College	Each	10,500.00
	Add for Back Stops	Each	1,250.00
.15	Volley Ball Courts - Court Size 30' x 60'	Each	9,300.00
	Minimum Use Size - 36' x 66'		
	Concrete - 4"	Each	8,200.00
	Asphalt - 1 1/2"	Each	6,500.00
	Add for Posts and Inserts	Pair	520.00
.16	Shuffleboard Courts - 6' x 52'		
	Concrete - 4"	Each	1,200.00
	Asphalt - 1 1/2"	Each	750.00
.17	Miniature Golf (9 - Hole) Bases - Concrete	Each	9,300.00
	Equipment and Carpet	Each	7,650.00
.18	Horse Shoe Courts - 10' x 40'	Each	520.00
	See Division 11 for Playground Equipment		
.19	Baseball and Softball Fields		
	45' Radius with Backstop	Each	12,000.00
	65' Radius with Backstop	Each	14,600.00

		UNIT	COST
0208.0 RECREATIONAL FACILITIES, Cont'd...			
.2 RECREATIONAL EQUIPMENT			
Climbers - Arch		Each	440.00
Dome		Each	515.00
Circular		Each	1,000.00
Catwalks		Each	640.00
Slides - Straight - 6'		Each	850.00
8'		Each	1,000.00
10'		Each	1,600.00
Spiral - 7'		Each	3,900.00
10'		Each	6,000.00
Spring Equipment - Single		Each	370.00
2 Unit		Each	565.00
4 Unit		Each	1,050.00
Swings - 2 Leg - 2 Seat - 8' High		Each	620.00
2 Leg - 4 Seat		Each	900.00
3 Leg - 3 Seat - 8' High		Each	695.00
6 Seat		Each	1,050.00
Add for 12' High		-	20%
Add for Nursery Seats		Each	250.00
Add for Saddle Seats		Each	570.00
Add for Tire Swings		Each	145.00
Teeter Totters		Each	925.00
Whirls - Small - 6'		Each	765.00
Large - 10'		Each	1,300.00
.3 SITE FURNISHINGS			
Benches - Steel Leg			
Stationary - w/Black Wood Slats - 8'		Each	380.00
w/Black Aluminum Slats - 8'		Each	515.00
Portable - 8'		Each	240.00
Player - No Back, Wood Seat - 8'		Each	200.00
No Back, Wood Seat - 14'		Each	320.00
Bike Racks			
Permanent - 10'		Each	515.00
20'		Each	800.00
Portable - 10'		Each	280.00
Bleachers, Elevated			
5 - Row x 15' Steel - Seats 50		Each	3,200.00
10 - Row x 15' Steel - Seats 100		Each	5,500.00
15 - Row x 15' Steel - Seats 150		Each	7,850.00
Litter Receptacles			
Pedestal - 24"		Each	290.00
36"		Each	360.00
Stoves		Each	340.00
Tables - Steel Frame - Wood Slat - 8'		Each	370.00
Aluminum Slat - 8'		Each	410.00
All Wood		Each	240.00
All Aluminum		Each	420.00
.4 SHELTERS		SqFt	14.00

		UNIT	COST	UNIT	COST
0209.0	**LANDSCAPING (L&M) (Laborers)**				
.1	SEEDING (.80 lb) - Machine	SqYd	.37	Acre	1,700.00
	Hand	SqYd	.85	Acre	4,200.00
	Add Fine Grading - Hand	SqYd	.75	Acre	3,600.00
	Add for 30 - Day Maintenance	-	-	Acre	1,200.00
	Deduct for Rye and Wild Grasses	-	-	Acre	550.00
.2	SODDING - Flat ($.80 per SqYd)	SqYd	2.30	-	-
	Slope (Pegged)	SqYd	2.80	-	-
	Add Fine Grading - Hand	SqYd	.85	-	-
	Add for 30 - Day Maintenance	SqYd	.38	-	-
	Remove and Haul Away Old Sod	SqYd	1.00		
.3	DISCING	SqYd	.80	Acre	4,000.00
.4	FERTILIZING	SqYd	.22	Acre	1000.00
				Ton	830.00
.5	TOP SOIL - 4" Hand - Only	SqYd	2.90	Acre	15,500.00
	Machine and Hand	SqYd	2.00	Acre	10,000.00
		CuYd	26.00	-	-
.6	TREES AND SHRUBS				
	Trees - 2"			Each	400.00
	3"			Each	560.00
	4"			Each	750.00
	5"			Each	940.00
	6"			Each	1,350.00
	Shrubs - Small - 1' to 3'			Each	19.00
	Large - 3' to 5'			Each	38.00
.7	BEDS				
	Wood Chips, Cedar, 2" - $12 CuYd			SqYd	2.50
	Mulch, Redwood Bark - $30 CuYd			SqYd	5.40
	Stone Aggregate - $15 CuYd			SqYd	3.00
	Planting Soil, 24" - $15 CuYd			SqYd	3.40
.8	WOOD CURBS AND WALLS				
	Curbs - 4" x 4" - Fir or Pine Treated			LnFt	2.20
	Cedar			LnFt	2.90
	6" x 6" - Fir or Pine Treated			LnFt	5.80
	Cedar			LnFt	7.25
	Walls - 4" x 4" - Fir or Pine, Incl. Dead Men			SqFt	6.10
	6" x 6" - Fir or Pine, Incl. Dead Men			SqFt	9.00
	Cedar, Including Dead Men			SqFt	13.80
.9	SPRINKLER SYSTEM, WATERED AREA			CSF	31.00
0210.0	**SOIL STABILIZATION**				
.1	CHEMICAL INTRUSION			CuFt	14.50
	Pressure Grout			Gal	4.60
.2	CONCRETE INTRUSION			CuFt	27.00
.3	VIBRO FLOTATION			-	-
0211.0	**SOIL TREATMENT**				
.1	PESTICIDE			SqFt	.24
.2	VEGETATION PREVENTATIVE			SqFt	.09
0212.0	**RAILROAD WORK**				
.1	NEW WORK			LnFt	125.00
.2	REPAIR WORK			LnFt	18.00
0213.0	**MARINE WORK**				
0214.0	**TUNNEL WORK**				

0201.0 DEMOLITION

.32   Selective Building Removals -

	UNIT	COST
No Cutting or Disposal Included (see 0201.34 & 0201.35)		
Concrete - Hand Work (see 0201.32 for Machine Work)		
8" Walls - Reinforced	SqFt	9.70
Non - Reinforced	SqFt	7.55
12" Walls - Reinforced	SqFt	19.00
12" Footings x 24" wide	LnFt	16.00
x 36" wide	LnFt	23.20
16" Footings x 24" wide	LnFt	29.60
6" Structural Slab - Reinforced	SqFt	6.20
8" Structural Slab - Reinforced	SqFt	6.90
4" Slab on Ground - Reinforced	SqFt	2.80
Non - Reinforced	SqFt	2.00
6" Slab on Ground - Reinforced	SqFt	3.25
Non - Reinforced	SqFt	2.35
Stairs - Reinforced	SqFt	10.50
Masonry - Hand Work		
4" Brick or Stone Walls	SqFt	2.00
4" Brick and 8" Backup Block or Tile	SqFt	3.60
4" Block or Tile Partitions	SqFt	1.85
6" Block or Tile Partitions	SqFt	1.95
8" Block or Tile Partitions	SqFt	2.30
12" Block or Tile Partitions	SqFt	2.95
Miscellaneous - Hand Work (Including Loading)		
Acoustical Ceilings - Attached (Including Iron)	SqFt	.67
Suspended (Including Grid)	SqFt	.40
Asbestos - Pipe	LnFt	49.50
Ceilings and Walls	SqFt	16.50
Columns and Beams	SqFt	41.30
Tile Flooring	SqFt	1.80
Cabinets and Tops	LnFt	9.30
Carpet	SqFt	.31
Ceramic and Quarry Tile	SqFt	1.30
Doors and Frames - Metal	Each	56.00
Wood	Each	49.50
Drywall Ceilings - Attached	SqFt	.83
Drywall on Wood or Metal Studs - 2 Sides	SqFt	.88
Paint Removal - Doors and Windows	SqFt	.88
Walls	SqFt	.62
Plaster Ceilings - Attached (Including Iron)	SqFt	1.25
on Wood or Metal Studs	SqFt	1.30
Roofing - Builtup	SqFt	1.10
Shingles - Asphalt and Wood	SqFt	.34
Terrazzo Flooring	SqFt	1.85
Vinyl Flooring	SqFt	.45
Wall Coverings	SqFt	.67
Windows	Each	37.00
Wood Flooring	SqFt	.41

.33   Site Removals (Including Loading)

	UNIT	COST
4" Concrete Walks - Labor Only (Non-Reinforced)	SqFt	1.95
Machine Only (Non-Reinforced)	SqFt	.64
6" Concrete Drives - Labor Only (Non-Reinforced)	SqFt	1.95
Machine Only (Reinforced)	SqFt	.94
6" x 18" Concrete Curb - Machine	LnFt	1.60
Curb and Gutter - Machine	LnFt	2.30
2" Asphalt - Machine	SqFt	.41
Fencing - 8' Hand	LnFt	1.95

		UNIT	COST
0202.0	**EARTHWORK**		
.1	GRADING - Hand - 4" - Site	SqFt	.20
	4" - Building	SqFt	.39
.2	EXCAVATION - Hand - Open - Soft (Sand)	CuYd	29.00
	Medium (Clay)	CuYd	38.00
	Hard (Shale)	CuYd	67.00
	Add for Trench or Pocket		15%
.3	BACK FILL - Hand - Not Compacted (Site Borrow)	CuYd	16.50
.4	BACK FILL - Hand - Compacted (Site Borrow)		
	12" Lifts - Building - No Machine	CuYd	23.00
	With Machine	CuYd	18.00
	18" Lifts - Building - No Machine	CuYd	21.00
	With Machine	CuYd	17.00
0205.0	**DRAINAGE**		
.4	BUILDING FOUNDATION DRAINAGE		
	4" Clay Pipe	LnFt	3.50
	4" Plastic Pipe - Perforated	LnFt	2.55
	6" Clay Pipe	LnFt	3.70
	6" Plastic Pipe - Perforated	LnFt	4.30
	Add for Porous Surround - 2' x 2'	LnFt	5.50
0206.0	**PAVEMENT, CURBS AND WALKS**		
.2	CURBS AND GUTTERS		
.21	Concrete - Cast in Place (Machine Placed)		
	Curb - 6" x 12"	LnFt	12.50
	6" x 18"	LnFt	15.00
	6" x 24"	LnFt	20.50
	6" x 30"	LnFt	22.00
	Curb and Gutter - 6" x 12"	LnFt	15.70
	6" x 18"	LnFt	19.60
	6" x 24"	LnFt	23.00
	Add for Hand Placed	LnFt	7.60
	Add for 2 #5 Reinf. Rods	LnFt	2.05
	Add for Curves and Radius Work	LnFt	40%
.22	Concrete Precast - 6" x 10" x 8"	LnFt	9.10
	6" x 9" x 8"	LnFt	8.40
.23	Bituminous - 6" x 8"	LnFt	2.75
.24	Granite - 6" x 16"	LnFt	32.00
.25	Timbers - Treated - 6" x 6"	LnFt	6.60
	6" x 8"	LnFt	9.00
.26	Plastic - 6" x 6"	LnFt	8.40
.3	WALKS		
	Bituminous - 1 1/2" with 4" Sand Base	SqFt	1.30
	2" with 4" Sand Base	SqFt	1.35
	Concrete - 4" - Broom Finish	SqFt	3.90
	5" - Broom Finish	SqFt	4.15
	6" - Broom Finish	SqFt	4.30
	Add for 6" x 6", 10 - 10 Mesh	SqFt	.30
	Add for 4" Sand Base	SqFt	.55
	Add for Exposed Aggregate	SqFt	1.30
	Crushed Rock - 4"	SqFt	.65
	Brick - 4" - with 2" Sand Cushion	SqFt	8.90
	4" - with 2" Mortar Setting Bed	SqFt	10.10
	Flagstone - 1¼" - with 4" Sand Cushion	SqFt	14.00
	1¼" - with 2" Mortar Setting Bed	SqFt	17.70
	Precast Block - 1" Colored with 4" Sand Cushion	SqFt	3.20
	2" Colored with 4" Sand Cushion	SqFt	3.80
	Wood - 2" Boards on 6" x 6" Timbers	SqFt	5.10
	2" Boards on 4" x 4" Timbers	SqFt	4.50
	Slate - 1¼" - with 2" Mortar Setting Bed	SqFt	18.00

DIVISION #3 - CONCRETE

Wage Rates (Including Fringes) & Location Modifiers
July 2001-2002

	Metropolitan Area	Placing Labor Rate		Cement Finisher Rate		Carpenter Rate		Wage Rate Location Modifier
1.	Akron	*	26.75	*	32.97	*	32.30	114
2.	Albany-Schenectady-Troy		25.26		31.73		28.48	102
3.	Atlanta		13.49		20.27		21.20	80
4.	Austin	**	15.34	**	20.06	**	21.18	74
5.	Baltimore		17.65		25.15		25.14	92
6.	Birmingham	**	17.15	**	22.19	**	21.17	75
7.	Boston		30.29		41.38	**	37.95	136
8.	Buffalo-Niagara Falls		30.75		36.73		36.98	128
9.	Charlotte	**	13.42	**	18.58	**	19.30	67
10.	Chicago-Gary	*	31.68	*	37.91	*	37.64	126
11.	Cincinnati		23.74		25.49		26.85	95
12.	Cleveland		29.10		33.35		32.60	117
13.	Columbus	*	22.63	*	27.28	*	27.66	99
14.	Dallas-Fort Worth	**	13.84	**	20.64	**	19.23	68
15.	Dayton	*	23.29	*	29.57	*	30.13	105
16.	Denver-Boulder		15.76		23.42		23.19	81
17.	Detroit		29.53		34.91		36.78	127
18.	Flint	*	24.36	*	29.46	*	29.24	100
19.	Grand Rapids		22.14		28.69		29.23	100
20.	Greensboro-West Salem	**	13.44	**	18.52	**	19.40	67
21.	Hartford-New Britain	*	27.50	*	34.90	*	32.97	116
22.	Houston	**	16.28	**	21.22	**	22.25	73
23.	Indianapolis		23.15		31.48		30.09	104
24.	Jacksonville	**	15.43	**	22.73	**	23.68	82
25.	Kansas City		23.32		28.93		28.64	98
26.	Los Angeles-Long Beach		29.07		34.38		31.79	122
27.	Louisville	**	18.63	**	24.21	**	23.46	82
28.	Memphis	**	17.03	**	20.59	**	19.97	87
29.	Miami	**	16.25	**	21.43	**	21.41	75
30.	Milwaukee	*	28.48	*	31.71	*	33.17	112
31.	Minneapolis-St. Paul		29.35		33.80		33.30	113
32.	Nashville	**	14.96	**	19.45	**	22.77	79
33.	New Orleans	**	14.04	**	17.44	**	20.81	73
34.	New York		41.00		50.05		53.62	193
35.	Norfolk-Portsmouth	**	15.07	**	19.47	**	22.38	79
36.	Oklahoma City	**	15.97	**	23.43	**	21.41	75
37.	Omaha-Council Bluffs		19.47		22.30		22.29	75
38.	Orlando	**	18.37	**	24.65	**	24.97	89
39.	Philadelphia	*	31.32	*	37.58	*	39.28	135
40.	Phoenix	**	17.39	**	21.88	**	23.88	95
41.	Pittsburgh	*	23.78	*	29.68	*	31.28	109
42.	Portland	*	27.60	*	33.00	*	32.29	113
43.	Providence-Pawtucket	*	26.63	*	31.79	*	32.40	114
44.	Richmond	**	14.96	**	19.01	**	22.34	78
45.	Rochester		26.03		33.71		33.67	116
46.	Sacramento	*	29.82	*	36.95	*	39.11	132
47.	St. Louis		29.15		33.13		34.26	117
48.	Salt Lake City-Ogden	**	16.67	**	23.37	**	22.16	70
49.	San Antonio	**	12.21	**	17.87	**	20.78	75
50.	San Diego	*	29.89	*	27.66	*	32.14	124
51.	San Francisco-Oakland-San Jose		29.85		32.45		39.11	130
52.	Seattle-Everett		29.56		36.04		33.61	116
53.	Springfield-Holyoke-Chicopee	*	30.92	*	30.07	*	31.45	110
54.	Syracuse	*	23.91	*	27.43	*	27.52	114
55.	Tampa-St. Petersburg	**	17.10	**	25.27	**	25.65	91
56.	Toledo	*	25.10	*	32.82	*	33.16	117
57.	Tulsa	**	15.93	**	23.43	**	21.83	75
58.	Tucson	**	20.02	**	27.35	**	27.06	95
59.	Washington D.C.	*	16.92	*	24.80	*	23.82	87
60.	Youngstown-Warren	*	26.67	*	30.55	*	29.28	107
	AVERAGE		**22.41**		**27.94**		**28.35**	

Note: See Division 5 for Reinforcing Steel Location Modifiers and Wage Rates
  *   Contract Not Settled - Wage Interpolated

Impact Ratio:	Placing	Labor 20%	Material 80%
	Finishing	Labor 92%	Material 8%
	Form Work	Labor 75%	Material 25%
	Reinforcing Steel	Labor 35%	Material 65%

  **   Non Signatory or Open Shop Rate

		PAGE
0301.0	**CONCRETE - Cast in Place (CSI 03300)**	**3-4**
.1	CONCRETE PLACING	3-4
.11	Footings	3-4
.12	Walls & Grade Beams	3-4
.13	Columns and Pedestals	3-4
.14	Beams	3-4
.15	Slabs (Structural)	3-4
.16	Stairs and Landings	3-4
.17	Curbs, Platforms and Miscellaneous Small Pours	3-4
.18	Slabs on Ground	3-4
.19	Slabs Over Decks or Lath	3-4
.20	Toppings	3-4
.2	CONCRETE FINISHING (INCLUDING SCREEDS)	3-5
.21	Rough Screeding	3-5
.22	Trowel Finishing	3-5
.23	Brush or Broom Finishing	3-5
.24	Float Finishing	3-5
.25	Special Finishing (Rub, Patch, Retard, Sandblasts, etc)	3-5
.3	SPECIALTIES	3-6
.31	Abrasives	3-6
.32	Admixtures	3-6
.33	Colors	3-6
.34	Curing	3-6
.35	Expansion and Control Joints	3-6
.36	Grouts	3-7
.37	Hardeners and Sealers	3-7
.38	Joint Sealers	3-7
.39	Moisture Proofing	3-7
.4	EQUIPMENT	3-7
.5	TESTS	3-7
0302.0	**FORMWORK (CSI 03100)**	**3-7**
.1	REMOVABLE FORMS (Expendable and Reusable)*	3-7
.11	Footings	3-7
.12	Walls and Grade Beams	3-7
.13	Columns	3-8
.14	Beams	3-8
.15	Slabs (Flat and Pan)	3-9
.16	Stairs and Landings	3-10
.17	Curbs and Platforms	3-10
.18	Edge Forms and Bulkheads for Slabs	3-10
.2	SHORING, SUPPORTS AND ACCESSORIES	3-10
.21	Horizontal Shoring (Beam and Joist)	3-10
.22	Vertical Shoring (Posts and Tubular Frames)	3-10
.23	Column Clamps and Steel Strapping	3-10
.24	Wall, Column and Beam Forms	3-10
.25	Accessories (Hangers, Tyloops, Lags, etc.)	3-11
.3	SPECIALTIES	3-11
.31	Nails, Wire and Ties	3-11
.32	Anchors and Inserts	3-11
.33	Chamfers, Drips, etc	3-11
.34	Stair Nosings and Treads	3-12
.35	Water Stops	3-12
.36	Divider Strips	3-12
.37	Tongue and Groove Joint Forms	3-12
.38	Bearing Pads and Shims	3-12
.4	EQUIPMENT	3-12

* See Section 0505 for Permanent Forms of Corrugated Deck and Steel Lath.

		PAGE
0303.0 REINFORCING STEEL (CSI 03200)		**3-13**
.1	BARS	3-13
.2	ACCESSORIES	3-13
.3	POST TENSION STEEL (Cable and Wire)	3-13
.4	WIRE FABRIC	3-14
0304.0 SPECIALTY PLACED CONCRETE		**3-14**
.1	LIFT SLAB	3-14
.2	POST TENSIONED OR STRESSED IN PLACE	3-14
.3	TILT UP PANELS	
.4	PNEUMATICALLY PLACED	3-14
0305.0 PRECAST PRESTRESSED CONCRETE		**3-14**
.1	BEAMS AND COLUMNS	3-14
.2	DECKS (Flat)	3-14
.3	SINGLE AND DOUBLE T'S	3-14
.4	WALLS	3-14
0306.0 PRECAST CONCRETE (CSI 03400)		**3-15**
.1	BEAMS AND COLUMNS	3-15
.2	DECKS	3-15
.3	PANELS (Wall and Facing Units)	3-15
.4	PLANK	3-15
.5	SPECIALTIES (Curbs, Copings, Sills, Stools, Medians, etc.)	3-15
0307.0 PRECAST CEMENTITIOUS PLANK (CSI 03500)		**3-15**
.1	EXPANDED MINERALS	3-15
.2	GYPSUM (CSI 03510)	3-15
.3	WOOD FIBRE (CSI 03530)	3-15
0308.0 POURED IN PLACE CEMENTITIOUS CONCRETE		**3-15**
.1	EXPANDED MINERALS	3-15
.2	EXPANDED MINERALS AND ASPHALT	3-15
.3	GYPCRETE	3-15
3A	**QUICK ESTIMATING**	**3-16**
		3-17
		3-18

0301.0 GENERAL CONCRETE WORK (S) (Cement Finishers & Laborers)

.1 CONCRETE PLACING & VIBRATING
(3500# Concrete w/5.5 sack cement)

		UNIT	LABOR	MATERIAL
.11	Footings - 1 1/2" Aggregate			
	Wall & Pad Types - Truck Chuted	CuYd	12.40	72.00
	Buggies	CuYd	16.00	72.00
	Crane	CuYd	17.00	72.00
.12	Wall & Grade Beams - 3/4" Aggregate			
	At Grade (to 8' deep) -Truck Chuted	CuYd	12.90	75.00
	Buggies	CuYd	17.50	75.00
	Crane	CuYd	17.50	75.00
	At Grade (over 8' deep,poured with trunks) - Truck Chuted	CuYd	13.90	75.00
	Buggies	CuYd	18.60	75.00
	Crane & Hoppers	CuYd	17.50	75.00
	Above Grade (deeper than 8') - Conveyors	CuYd	12.90	75.00
	Ramp & Buggies	CuYd	18.60	75.00
	Crane & Hoppers	CuYd	17.50	75.00
	Climbing Crane	CuYd	16.40	75.00
	Pumping	CuYd	13.90	75.00
.13	Column & Pedestals - 3/4" Aggregate - Buggies	CuYd	20.60	75.00
	Crane	CuYd	17.50	75.00
	Tower & Buggies	CuYd	19.60	75.00
	Fork Lift	CuYd	36.00	75.00
	Climbing Crane	CuYd	18.60	75.00
	Pumping	CuYd	15.00	75.00
.14	Beams - 3/4" Aggregate - Buggies	CuYd	19.60	75.00
	Crane	CuYd	17.50	75.00
	Tower & Buggies	CuYd	19.60	75.00
	Fork Lift	CuYd	36.00	75.00
	Climbing Crane	CuYd	17.50	75.00
	Pumping	CuYd	15.50	75.00
.15	Slabs - 6" Structure - 3/4" Aggregate - Buggies	CuYd	16.60	75.00
	Crane	CuYd	17.50	75.00
	Climbing Crane	CuYd	17.50	75.00
	Pumping	CuYd	15.00	75.00
.16	Stairs & Landings -			
	Structural - 3/4" Aggregate - Buggies	CuYd	19.60	75.00
	Truck Chuted	CuYd	15.50	75.00
	Climbing Crane	CuYd	17.50	75.00
	Pumping	CuYd	15.50	75.00
	Pan Filled - 1/2" Ready Mixed	CuYd	34.00	80.00
	Hand Mixed (Dry)	CuYd	53.50	67.00
.17	Curbs, Platforms & Misc.Small Pours - 1/2" Aggregate Buggies	CuYd	24.65	80.00
	Fork Lift	CuYd	34.00	80.00
.18	Slabs on Ground - 3/4" Aggregate - Truck Chuted	CuYd	13.40	75.00
	Buggies	CuYd	17.50	75.00
	Crane	CuYd	16.60	75.00
	Conveyors	CuYd	13.90	75.00
.19	Slabs-Over Metal Decks or Lath - 1/2" Aggregate Buggies	CuYd	20.60	80.00
	Tower & Buggies	CuYd	21.70	80.00
	Crane	CuYd	18.55	80.00
.20	Toppings - 1/2" Aggregate - Buggies	CuYd	19.60	80.00
	Towers & Buggies	CuYd	21.70	80.00

Conveying Equipment and Operator not included in above costs. See Page 3-5. Also other variations on Page 3-5.

0301.0	GENERAL CONCRETE WORK, Cont'd...	UNIT	LABOR	MATERIAL
	Additions to Concrete Placing Work:			
	Add for Ea 500# Concrete above 3500# Concrete	CuYd	-	2.00
	Add or deduct for Sack Cement	CuYd	-	5.00
	Add for High Carbon Concrete	CuYd	-	18.00
	Add for High Early Cement Concrete	CuYd	-	7.50
	Deduct for 1 1/2" Aggregate	CuYd	-	1.50
	Add for 3/8" or 1/2" Aggregate	CuYd	-	3.00
	Add for Lightweight Aggregate (4000#)	CuYd	2.55	15.00
	Add for Heavyweight Aggregate (Granite) (4000#)	CuYd	5.15	10.00
	Deduct for Fly Ash Concrete (3000#)	CuYd	-	1.00
	Add for Exposed Aggregate Mix	CuYd	-	8.00
	Add for Fibre Reinforcement Mix	CuYd	3.10	8.00
	Add for Less than 6 CuYd Delivery	Load	-	45.00
	Add for Hauls beyond 15 Miles	CuYd Mi	-	1.00
	Add for Time after 7 Minutes per Yard	Minute	-	.75
	Add for Conveying Equipment and Operators:			
	Towers (Approx. 30 CuYd/ Hour)	CuYd	-	5.75
	Mobile Cranes (Approx. 25 CuYd/ Hour)	CuYd	-	6.75
	Tower Cranes (Approx. 20 CuYd/ Hour)	CuYd	-	7.25
	Fork Lifts (Approx. 5 CuYd/ Hour)	CuYd	-	6.75
	Conveyors (Approx. 50 CuYd/ Hour)	CuYd	-	7.25
	Pumping (Approx. 35 CuYd/ Hour)	CuYd	-	7.25
	Add for Floors Above Grade - per Floor Placing	CuYd	1.03	-
	Add for Winter Work:			
	Productivity Loss	CuYd	4.15	-
	Heating Water and Aggregate	CuYd	-	4.00
	Calcium Chloride 1%	CuYd	-	2.00
	Insulation Blanket - Slabs (5 uses)	SqFt	.07	.12
	Walls & Beams (3 uses)	SqFt	.09	.10
	Heaters: Without Operators (Including Fuel - Floor Area)	CuYd	.36	1.40
		or SqFt	.04	.11
	Enclosures (3 uses)- Wall Area	CuYd	.83	.35
		or SqFt	.07	.15
.2	FINISHING (Including Screeds and Bulkheads to 6")			
.21	Rough Screeding - Slabs	SqFt	.27	.06
	Stairs	SqFt	.47	.06
.22	Trowel Finishing - Slabs on Ground	SqFt	.41	.06
	Solid and Pan Slabs	SqFt	.43	.06
	Slab over Corrugated	SqFt	.49	.06
	Topping Slabs	SqFt	.43	.06
	Stairs	SqFt	.75	.06
	Curbs and Bases	SqFt	.75	.06
.23	Broom Finishing - Slabs on Ground	SqFt	.36	.06
	Solid and Pan Slabs	SqFt	.41	.06
	Slab over Corrugated	SqFt	.49	.06
	Topping Slabs	SqFt	.41	.06
	Stairs	SqFt	.75	.06
.24	Float Finish - Slabs on Ground	SqFt	.34	.06
	Solid and Pan Slabs	SqFt	.38	.06
	Slab over Corrugated	SqFt	.41	.06
	Stairs	SqFt	.67	.06
.25	Special Finishing			
.251	Patch Walls, 2 Sides Tie Holes & Honeycomb	SqFt	.14	.06
.252	Rub Walls, 1 Side Carborundum for Fins	SqFt	.08	.03
	With Burlap and Grout	SqFt	.43	.07
.253	Level & Top Floors - Trowel Finish 1"	SqFt	.75	.33
	With Epoxy 1/4"	SqFt	1.50	4.90
.254	Exposed Aggregate Slabs - Washed	SqFt	.57	.04
	Retardant	SqFt	.19	.09
	Add for - Seeding	SqFt	.28	.10
	Add to All Finishing Items Above			
	Winter Production Loss and Cost	SqFt	.14	.04
	Sloped Work	SqFt	.11	.03
	Heavyweight Aggregates	SqFt	.14	.04

	UNIT	LABOR	MATERIAL
0301.0 GENERAL CONCRETE WORK, Cont'd...			
.255 Bushhammer - Green Concrete	SqFt	.98	.07
Cured Concrete	SqFt	1.50	.09
Sand Blast - Light Penetration	SqFt	.72	.10
Heavy Penetration	SqFt	1.25	.20
.3 SPECIALTIES			
.31 Abrasives - Non-Slip			
Alo-Grit (.86 lb & 1/4 lb/SqFt)	SqFt	.18	.22
Carborundum Grits (1.30 lb & 1/4 lb/SqFt)	SqFt	.18	.35
Strips 3/8" x 1/4"	LnFt	.66	1.30
Epoxy Coating (50.00 gal) & 50 SqFt/Gal	SqFt	.26	1.40
.32 Admixtures (per CuYd Concrete)			
Accelerator (.06 oz)	CuYd	-	1.35
Air Entraining (.04 oz)	CuYd	-	.75
Densifiers (.07 oz)	CuYd	-	1.70
Retarders (.05 oz)	CuYd	-	.75
Water Reducing (.06 oz)	CuYd	-	1.95
.33 Colors			
Dust on Type:			
Black, Brown & Red (.50 lb) 50#/100 SqFt	SqFt	.18	.33
Green and Blue (.60 lb) 50#/100 SqFt	SqFt	.18	.41
Integral (Top 1"):			
Black, Brown & Red (1.50 lb) 27#/100 SqFt	SqFt	.33	.57
Green and Blue (4.25 lb) 27#/100 SqFt	SqFt	.33	1.28
Full Thickness			
Black, Brown & Red	CuYd	20.50	112.00
Green and Blue	CuYd	20.50	150.00
.34 Curing			
Curing Compounds			
Resin (10.00 gal) 200 SqFt/gal	SqFt	.07	.05
Hydrocide Res.Base (15.00 gal) 200 SqFt/gal	SqFt	.06	.07
Rubber Base (8.00 gal) 200 SqFt/gal	SqFt	.06	.04
Wax Base (3.00 gal) 200 SqFt/gal	SqFt	.06	.02
Asphalt Base (4.00 gal) 200 SqFt/gal	SqFt	.06	.02
Paper	SqFt	.07	.06
Polyethylene - 4 mil	SqFt	.07	.04
Water	SqFt	.07	.01
Burlap	SqFt	.08	.06
Curing and Sealing	SqFt	.09	.06
.35 Expansion and Control Joints			
Asphalt - Fibre - 1/2" x 4"	LnFt	.31	.21
1/2" x 6"	LnFt	.32	.31
1/2" x 8"	LnFt	.38	.41
Polyethylene Foam - 1/2" x 4"	LnFt	.31	.36
1/2" x 6"	LnFt	.33	.46
1/2" x 8"	LnFt	.38	.57
Sponge Rubber - 1/2" x 4"	LnFt	.31	1.90
1/2" x 6"	LnFt	.33	2.95
1/2" x 8"	LnFt	.38	3.95
Paper - Fibre (No Oil) - 1/2" x 4"	LnFt	.31	.26
1/2" x 6"	LnFt	.33	.35
1/2" x 8"	LnFt	.38	.44
Add for Cap	LnFt	.17	.27
Add for 3/4" Thickness	LnFt	-	30%
Add for 1" Thickness	LnFt	-	80%

See 0413.5 for Other Expansion Joint Costs.
See 0410.0 Mortars for Cement Prices.

0301.0 GENERAL CONCRETE WORK, Cont'd...

		UNIT	LABOR	MATERIAL
.36	Grouts - Non-Shrink			
	Iron Oxide (.45/lb)			
	Hand Mixed (1 Cem: 1 Sand: 1 Iron Oxide)	CuFt	5.65	25.50
	Per 1"	SqFt	.72	2.05
	Premixed (.45/lb)	CuFt	2.05	37.00
	Aluminum Oxide (.45/lb)	CuFt	4.10	11.00
	Non-Metallic, Premixed (.35/lb)	CuFt	3.05	27.00
.37	Hardeners and Sealers			
	Acrylic Sealer (7.50 gal) 300 SqFt/gal	SqFt	.09	.05
	Epoxy Sealer (25.00 gal) 300 SqFt/gal	SqFt	.10	.13
	Urethane Sealer (9.50 gal) 400 SqFt/gal	SqFt	.10	.05
	Liquid Hardeners (4.85 gal) 200 SqFt/gal	SqFt	.08	.05
.38	Joint Sealers			
	Rubber Asphalt			
	Hot 1/2" x 1/2" Joint (.70 lb)	LnFt	.50	.45
	Cold 1/2" x 1/2" Joint (.54 lb)	LnFt	.35	.52
	Epoxy (20.00 Qt)	LnFt	.50	1.35
.39	Moisture Proofing (Loose Laid for Slabs)			
	Polyethylene 4 Mil	SqFt	.07	.04
	6 Mil	SqFt	.07	.05
	Asphalted Paper	SqFt	.07	.06
.4	EQUIPMENT (Conv., Finishing, Vibrating, etc.)			
	See Divisions I-6A and I-7A			
.5	TESTS			
	Cylinder - 6" x 12" (7-day and 28-day)	Ea	-	37.00
	Add per Pickup	Ea	-	36.00

0302.0 CONCRETE FORM WORK (S) (Carpenters)
(Lumber @ 530.00 MBF & 3/4" BB Plywood @ 1.10 SqFt)
(Costs Include Erection, Stripping, Cleaning, Oiling)

		UNIT	LABOR	MATERIAL
.1	REMOVABLE FORMS (Expendable and Reusable)			
.11	Footings (2-1/2 BdFt/SqFt and 3 Uses)			
	Wall Type - Constant Elevation	SqFt	1.95	.55
	Pad Type	SqFt	2.05	.55
	Add for Volume Elevations and Changes	SqFt	.15	-
	Add for Deep Foundation (8' or more)	SqFt	.14	-
	Add for Each Foot Form Deeper than 12"	SqFt	.09	.04
	Add for Keyway	LnFt	.39	.11
.12	Walls & Grade Beams (2-1/2 BdFt/SqFt & 3 Uses)			
	Straight Walls - 4' High	SqFt	2.00	.66
	Add for Each Foot Higher than 4'	SqFt	.04	.04
	Add for Walls Above Grade	SqFt	.06	-
	Add for Deep Foundation (8' or more)	SqFt	.10	-
	Add for Pilastered Wall, 24' O.C.Total Area	SqFt	.09	.02
	Add for Pilastered Wall, Pilasters Only	SqFt	3.90	.56
	Add for Openings - Opening Area	SqFt	2.60	.67
	Add for Retaining and Battered Type Wall	SqFt	.77	.09
	Add for Curved Wall	SqFt	1.55	.12
	Add for Brick Ledge	LnFt	1.40	.13
	Add for Parapet Wall - Hung	SqFt	2.30	.23
	Add for Pit or Small Trench Walls	SqFt	1.45	.08
	Add for Lined Forms - Hardwood Panels	SqFt	.34	.42
	Gang Formed Walls - Make Up	SqFt	3.60	5.60
	Move	SqFt	.83	.16
	3/4" B-B Plywood - Avg - 5 Ply - 8 Uses @ 1.10	SqFt	-	-
	7 Ply - 8 Uses @ 1.25	SqFt	-	-
	M.D.O. Plywood - Avg. - 20 Uses @ 1.45	SqFt		

0302.0 CONCRETE FORM WORK, Cont'd...

	UNIT	LABOR	MATERIAL
.13 Columns & Pedestals			
Sq. & Rectangular (2 1/2 BdFt/ SqFt & 3 Uses)			
8" x 8"	SqFt	3.85	.78
12" x 12"	SqFt	3.80	.79
16" x 16"	SqFt	3.75	.80
20" x 20"	SqFt	3.60	.81
24" x 24"	SqFt	3.65	.82
Add per Foot - Work over 10' Floor Heights	SqFt	.31	-
Add per Floor above 20' Above Grade	SqFt	.12	-
Deduct for Ganged Formed (8 Uses)	SqFt	1.05	-
Round - Steel or Fiberglass (Rent and 3 Uses)			
12"	LnFt	5.90	3.80
16"	LnFt	5.75	4.20
20"	LnFt	6.80	5.00
24"	LnFt	9.90	5.70
30"	LnFt	12.90	7.00
36"	LnFt	16.10	8.80
Add per Foot - Work over 10" Floor Heights	LnFt	.31	.40
Round-Fibre (6" to 48" x 18' available)			
8" Cut Length Prices Used	LnFt	4.15	2.15
12"	LnFt	4.45	4.20
16"	LnFt	4.75	6.00
20"	LnFt	5.90	9.00
24"	LnFt	8.00	12.00
30"	LnFt	9.30	15.00
36"	LnFt	11.40	17.00
Add for Less than 100 Feet - One Size	LnFt	-	20%
Add for Seamless Fibre	LnFt	-	15%
Add for Conical Heads to Steel or Fibre	Each	83.00	50.00
Add for Beam Fittings or Other Openings	Each	93.00	55.00
.14 Beams (12" Floor Heights and 3 Uses Lumber)			
Spandrel Beams (Includes Shoring)			
12" x 48" (4 1/2 BdFt/ SqFt)	SqFt	4.25	.97
12" x 42" (4 3/4 BdFt/ SqFt)	SqFt	4.30	.99
12" x 36" (5 BdFt/ SqFt)	SqFt	4.40	.96
12" x 30" (5 1/4 BdFt/ SqFt)	SqFt	4.55	1.05
8" x 42" (4 3/4 BdFt/ SqFt)	SqFt	4.55	1.00
8" x 36" (5 1/4 BdFt/ SqFt)	SqFt	4.65	1.05
Add for Decks and Safety Rail	SqFt	.88	.35
Interior Beams			
16" x 30" (4 BdFt/ SqFt)	SqFt	4.25	.95
16" x 24" (4 1/2 BdFt/ SqFt)	SqFt	4.30	.98
12" x 30" (4 3/4 BdFt/ SqFt)	SqFt	4.35	1.00
12" x 24" (5 1/4 BdFt/ SqFt)	SqFt	4.30	1.07
12" x 16" (6 1/2 BdFt/ SqFt)	SqFt	4.75	1.20
8" x 24" (5 1/4 BdFt/ SqFt)	SqFt	4.65	1.10
8" x 16" (6 1/2 BdFt/ SqFt)	SqFt	4.85	1.20
Add for Beams Carrying Horizontal Shoring	SqFt	.83	.28
Add per Foot for Floor Heights over 10'	SqFt	.41	.09
Add per Floor over 20' Above Grade	SqFt	.10	.08
Add for Splayed Beams	SqFt	1.45	.16
Add for Inverted Beams	SqFt	1.85	.17
Add for Mud Sills (3 Uses)	SqFt	.67	.35

0302.0 CONCRETE FORM WORK, Cont'd...

	UNIT	LABOR	MATL
.15 Slabs (12' Floor Heights and 3 Uses Lumber)			
.151 Solid Slabs (Includes Shoring) (Metal Adj. Beams or Joists)			
Horizontal Shoring Method			
To 5" thick (2' OC x 10' Span) 2.5 BdFt/ SqFt	SqFt	1.70	.63
5" - 8" thick (2' OC x 15' Span) 2.8 BdFt/ SqFt	SqFt	1.75	.70
8" - 11" thick (2' OC x 20' Span) 3.1 BdFt/ SqFt	SqFt	2.00	.75
Add for Heights above 10' per foot	SqFt	.21	.07
Vertical Shoring Method			
(Wood or Metal Posts, Purlins & Joists)			
To 5" thick - 2.7 BdFt/SqFt - 4' x 5'6" OC	SqFt	1.75	.68
5" - 8" thick - 3.0 BdFt/SqFt - 4' x 5'0" OC	SqFt	1.95	.73
8" - 11" thick - 3.3 BdFt/SqFt - 4' x 4'6" OC	SqFt	3.30	.77
Add for Heights above 10' per foot	SqFt	.21	.06
Add for Adjustable Hardware	SqFt	-	.02
Add for Cantilevered Slabs	SqFt	1.15	.14
Add for Drop Panels - No Edge included	SqFt	.21	.10
Flying Form method (8 Uses)			
To 5" Thick	SqFt	1.45	.57
5" - 8" Thick	SqFt	1.50	.58
8" - 11" Thick	SqFt	1.60	.59
Add for Heights above 8' per foot	SqFt	.19	.07
Add for 2 Uses	SqFt	.17	.24
Add for 4 Uses	SqFt	.13	.18
.152 Pan Slabs - Cost same as solid slabs above plus pan forming costs below.			
Based on 3 Uses and lease of approximately 4,000 Square Feet			
20" Pan 8" + 2"	SqFt	.79	.68
10" + 2"	SqFt	.81	.70
12" + 2"	SqFt	.84	.72
14" + 2"	SqFt	.89	.75
30" Pan 8" + 2 1/2"	SqFt	.76	.65
10"+2 1/2"	SqFt	.78	.67
12"+2 1/2"	SqFt	.81	.70
14"+2 1/2"	SqFt	.84	.72
19" x 19" Dome Pan - 24" x 24" Joist Center			
6" + 2"	SqFt	.76	.67
8" + 2"	SqFt	.79	.69
10" + 2"	SqFt	.81	.70
12" + 2"	SqFt	.86	.73
30" x 30" Dome Pan - 36" x 36" Joist Center			
10" + 2 1/2"	SqFt	.86	.74
12" + 2 1/2"	SqFt	.91	.76
14" + 2 1/2"	SqFt	.97	.80
Add for 2 Uses	SqFt	.06	.10
Add for 1 Use	SqFt	.07	.20
Deduct for 4 Uses	SqFt	.06	.06
Deduct for 5 Uses	SqFt	.05	.07
Pan Slabs: Based on 5 Uses & Lease of 4,000 SqFt			
40" x 40" Dome Pan - 16" + 3"	SqFt	1.05	.98
48" x 48" Joist Center - 18" + 3"	SqFt	1.15	1.18
Deduct for 6 Uses	SqFt	.06	.09
Deduct for 7 Uses	SqFt	.05	.10
Add to above units if subcontracted		30%	10%

0302.0 CONCRETE FORM WORK, Cont'd...

		UNIT	LABOR	MATL
.16	Stairs and Landing (3 BdFt and 2 Uses)			
	Structural - Contact Area of			
	Soffits, Stringers and Risers	SqFt	3.60	1.13
	On Ground - Contact Area of			
	Stringers and Risers	SqFt	4.70	1.18
.17	Curbs & Platforms (2 BdFt and 2 Uses)	SqFt	3.35	.67
.18	Edge Forms/Bulkheads for Slabs (2 BdFt, 2 Uses)			
	Floor Supported	SqFt	3.30	.68
	Structural	SqFt	4.50	.75

.2	SHORING, SUPPORTS AND ACCESSORIES			
	(All items in this section are included in SqFt			
	costs of Section 0302.1. This section is for			
	cost-keeping & limited analyzing of variables.)			

		UNIT		MATL
.21	Horizontal Shoring			
	10' Span - Monthly Rent Cost	Each		1.65
	SqFt Contact Area	SqFt		.35
	15' Span - Monthly Rent Cost	Each		6.25
	SqFt Contact Area	SqFt		.46
	20' Span - Monthly Rent Cost	Each		7.40
	SqFt Contact Area	SqFt		.62
.22	Vertical Shoring (4' OC - 10' High) - Average	SqFt		.26
	Wood Posts with Ellis Shore Hardware			
	Single - W/T Head - New	Each		11.80
	LnFt Beam - 6 Uses	LnFt		.60
	Double - W/T Head & Bracing - New	Each		23.50
	LnFt Beam - 6 Uses	LnFt		1.00
	Metal Posts - Single - Monthly Rent Cost	Each		3.35
	LnFt Beam - 2 Uses per Month	LnFt		.75
	Double - W/T Head & Bracing	LnFt		6.35
	LnFt Beam - 2 Uses per Month	LnFt		1.45
	Tubular Frames - to 10'			
	Mo. Cost - Incl. Head & Base - 10,000#	Each		13.80
	LnFt - Average - 2 Uses per Month	LnFt		1.15
	Deduct for 6,000#	Each		25%
.23	Column Clamps and Steel Strapping			
	Clamps - Monthly Rent Cost - Steel	Each		4.00
	Lock Fast	Each		7.60
	Gang Form	Each		6.70
	Contact Area - 2 Uses per Month	SqFt.		33
	Strapping - 3/4" x .025 (1.10 lb)	LnFt		.23
	3/4" x .023 (1.00 lb)	LnFt		.22
	Contact Area of Forms	SqFt		.14
.24	Wall and Beam Brackets			
	Brackets (2' OC) - Monthly Rent Cost	Each		.55
	New Cost	Each		5.60
	Contact Area - 4 Uses per Month	SqFt		.13
.25	Accessories - Beam Hangers			
	Tyloops, Lags, Washers & Bolts - Heavy	LnFt		.85
	Light	LnFt		.55

0302.0 CONCRETE FORM WORK, Cont'd...

		UNIT	LABOR	MATERIAL
.3	SPECIALTIES			
.31	Nails, Wire and Ties			
	Nails 6 d common (50# box)	CWT	-	54.00
	8 d common	CWT	-	46.00
	16 d common	CWT	-	44.00
	8 d box	CWT	-	78.00
	16 d box	CWT	-	76.00
	9 ga concrete nails (3/4" to 3")	CWT	-	130.00
	Add for Coated Nails	CWT	-	60%
	Wire - Black Annealed 9 ga (100# Coil)	CWT	-	65.00
	16 ga (100# Coil)	CWT	-	62.00
	Average Nail and Wire Cost for Formwork	BdFt	-	.015
		or SqFt	-	.03
	Ties - Breakback -8" 3000# 5" ends	Each	-	.60
	12" 3000# 5" ends	Each	-	.70
	16" 3000# 5" ends	Each	-	.75
	Add for 5000#	Each	-	.13
	Add for 8" ends	Each	-	.07
	Average Tie Cost for Formwork (2' OC)			
	Wall Area	SqFt	-	.33
	Contact Area Form	SqFt	-	.17
	Gang Form Contact Area	SqFt	-	.18
	Coil Bolts - 8"	Each	-	1.20
	12"	Each	-	1.40
.32	Anchors and Inserts (including Layout)			
	Anchor Bolts (Concrete) -			
	1/2" x 12"	Each	.95	.68
	5/8" x 12"	Each	1.05	1.75
	3/4" x 12"	Each	1.20	2.55
	Ceiling Type Inserts			
	Adjustable - 1/2" Bolt Size x 3'	Each	.75	2.00
	5/8" Bolt Size x 3'	Each	.80	2.05
	3/4" Bolt Size x 3'	Each	.85	2.40
	Continuous Slotted	Each	.65	1.60
	Threaded - 1/2" Bolt Size	Each	.75	3.35
	5/8" Bolt Size	Each	.80	5.50
	3/4" Bolt Size	Each	.85	6.50
	Hanger Wire Type - Drive In	Each	.65	.16
	Shell	Each	.65	.20
	Ferrule Type - 1/2"	Each	.70	.95
	5/8"	Each	.75	1.15
	3/4"	Each	.80	1.60
	T-Hanger - 4"	Each	.70	.85
	10"	Each	.70	1.30
	14"	Each	.85	1.70
	Wall and Beam Type Insert			
	Dovetail Slot - 24 ga (No Anchors)	LnFt	.28	36
	22 ga (No Anchors)	LnFt	.35	.70
	Flashing Reglects	LnFt	.45	.56
	Shelf Angle Inserts - Add for Bolts,			
	Washers, and Nuts -5/8"	Each	.75	2.60
	3/4"	Each	.80	3.25

0302.0	CONCRETE FORM WORK, Cont'd...	UNIT	LABOR	MATERIAL
	Bolts - 5/8" x 2"	Each	.67	1.05
	3/4" x 2"	Each	.68	1.65
	3/4" x 3"	Each	.70	2.30
	Nuts - 5/8"	Each	.25	.19
	3/4"	Each	.28	.30
	Washers	Each	.06	.22
	Threaded Rod - 3/8" x 36" (Galv)	Each	.87	1.15
	1/2" x 36" (Galv)	Each	.98	2.00
	5/8" x 36" (Galv)	Each	1.15	3.15
.33	Chamfers, Drips and Reveals (1 Use)			
	Chamfers - 3/4" Metal	LnFt	.23	.42
	Plastic with Tail	LnFt	.25	.48
	without Tail	LnFt	.25	.40
	Wood	LnFt	.22	.14
	Reveals - 3/4" Plastic	LnFt	.33	1.20
	Drips	LnFt	.29	.35
.34	Stair Nosings and Treads			
	Nosings, Curb Bar - Steel - Galvanized	LnFt	1.05	4.00
	Treads 3" x 1/2" - Cast Iron	LnFt	1.20	6.70
	Extruded Aluminum	LnFt	1.20	5.90
	Cast Aluminum	LnFt	1.20	7.20
.35	Water Stops			
	Polyvinyl Chloride - 6" x 3/16"	LnFt	.69	1.20
	6" Centerbulb	LnFt	.74	2.35
	9" x 3/16"	LnFt	.80	1.75
	9" Centerbulb	LnFt	.87	3.00
	Add for 3/8"	LnFt	.14	50%
	Rubber (Neoprene) - 6" Centerbulb	LnFt	.77	7.70
	9" Centerbulb	LnFt	.87	14.50
	Add per Splice	Each	6.50	-
.36	Divider Strips - White Metal, 16 ga x 1 1/4"	LnFt	.59	.82
	1/2" x 1 1/4"	LnFt	.60	.88
	3/16" x 1 1/4"	LnFt	.60	.88
	1/4" x 1 1/4"	LnFt	.60	1.40
	Add for Brass 1/8"	LnFt	.10	.78
.37	Tongue and Groove Joint Forms			
	Asphalt - 3 1/2" x 1/8"	LnFt	.64	.67
	5 1/2" x 1/8"	LnFt	.70	.82
	Metal - 3 1/2"	LnFt	.64	.56
	5 1/2"	LnFt	.70	.72
	Wood - 3 1/2"	LnFt	.64	.22
	5 1/2"	LnFt	.70	.32
	Plastic - 3 1/2"	LnFt	.64	.80
	Add for Stakes	Each	.62	.50
.38	Bearing Pads and Shims			
	Vinyl - 1/8" x 6" x 24"	SqFt	.72	1.85
	1/4" x 6" x 24"	SqFt	.83	3.70
	Neoprene - 1/8" x 36" (70 Duro)	SqFt	.62	4.80
	1/4" x 36"	SqFt	.67	8.50
.4	EQUIPMENT			

See Pages 1 - 6A and 1 - 7A. Average 5% Labor Cost.

Add to Labor Formwork Costs Above for Winter Construction 10% to 25%

0303.0 REINFORCING STEEL (S&L) (Ironworkers)

	UNIT	LABOR Subcont	MATERIAL Plain	Epoxy
.1 BARS (Based on Local Trucking)				
1/4" to 5/8" Large Size Job (100 Tons Up)	Ton	410	600	1,400
Medium Job (20 - 99 Tons Up)	Ton	460	640	1,460
Small Size Job (5 - 19 Tons)	Ton	480	690	1,570
3/4" to 1 1/4" Large Job (100 Tons Up)	Ton	420	600	1,380
Medium Job (10 - 99 Tons)	Ton	430	610	1,400
Small Size Job (5 - 19 Tons)	Ton	450	650	1,480
1 1/2" to 2" Average	Ton	440	630	1,440
Add For Galvanizing	Ton	52	430	-
Add for Bending - Light #2 to #5	Ton	-	230	-
Heavy #6 and Larger	Ton	-	115	-
Add for Size Extras - Light #2 to #4	Ton	-	140	-
Heavy #3 and Larger	Ton	-	105	-
Add for Spirals - Shop Assembled - Hot Rolled	Ton	-	160	-
Cold Rolled	Ton	-	225	-
Unassembled - Hot Rolled	Ton	-	135	-
Cold Rolled	Ton	-	215	-
Add for Splicing: #11 Bars - Buttweld	Each	29.00	2.90	-
#14 Bars - Buttweld	Each	39.00	4.20	-
#11 Bars - Mech. Buttweld	Each	40.00	10.80	-
#14 Bars - Mech. Buttweld	Each	46.00	11.85	-

BARS (Based on Local Trucking)		LABOR	MATERIAL	
Tables of Weights & Prices by LnFt -	Ton	-	800.00	
Small job from warehouse stock for local delivery	Lb	-	.40	
# 2 Bar 1/4" or .250 .167 lb	LnFt	.11	.11	
# 3 Bar 3/8" or .375 .376 lb	LnFt	.14	.17	
# 4 Bar 1/2" or .500 .668 lb	LnFt	.19	.26	
# 5 Bar 5/8" or .625 1.043 lb	LnFt	.28	.40	
# 6 Bar 3/4" or .750 1.502 lb	LnFt	.38	.55	
# 7 Bar 7/8" or .875 2.044 lb				
# 8 Bar 1" or 1.000 2.670 lb				
# 9 Bar 1 1/8" or 1.128 3.400 lb				
#10 Bar 1 1/4" or 1.270 4.303 lb				
#11 Bar 1 3/8" or 1.410 5.313 lb				
#14 Bar 1 1/2" or 1.693 7.650 lb				
#18 Bar 2 1/4" or 2.257 13.600 lb				

.2 ACCESSORIES
Additional sizes, but below are close to standard
and are included in tonnage costs above

	UNIT	LABOR	MATERIAL	
1/4" to 1 1/2" Slab Bolsters	LnFt	-	.27	
1/2" to 2" Beam Bolsters - Upper & Lower	LnFt	-	.55	
2" to 3" Beam Bolsters - Upper & Lower	LnFt	-	.85	
3" High Chairs - Continuous	LnFt	-	.40	
3/4" x 4" to 6" Joist Chairs	Each	-	.35	
1/2" Dowel Bar Tubes - Plastic and Metal	Each	.41	.25	
Add for Galvanized Accessories	-	-	.15	
Add for Plastic Accessories	-	-	.18	

.3 POST TENSION STEEL - PRESTRESSED IN FIELD

	UNIT	LABOR	MATERIAL	
Ungrouted - 100 kips	Lb	.70	1.20	
200 kips	Lb	.65	1.25	
Deduct for Added 100 kips	Lb	.02	-	
Add for Chairs	Lb	-	.12	

0303.0 REINFORCING STEEL, Cont'd...

	UNIT	LABOR	MATERIAL
.4 WIRE FABRIC (750 SqFt per Roll)			
6" x 6" x 10/10 (W 1.4)	SqFt	.11	.07
6" x 6" x 8/8 (W 2.1)	SqFt	.13	.11
6" x 6" x 6/6 (W 2.9)	SqFt	.16	.14
6" x 6" x 4/4 (W 4.0)	SqFt	.19	.17
4" x 4" x 4/4 (W 4.0)	SqFt	.24	.23
4" x 6" x 8/8 (W 2.0)	SqFt	.15	.14
Add for Epoxy Coated (120 SqFt Sheet)	SqFt	.02	.19

0304.0 SPECIALTY PLACED CONCRETE (No Fees)

.1 LIFT SLAB CONSTRUCTION

	UNIT	MATERIAL
Concrete - 6" Slab	SqFt	3.55
Formwork	SqFt	1.05
Reinforcing Steel	SqFt	1.20
Lifting	SqFt	1.20
Brace and Assemble	SqFt	.75
TOTAL		7.75

.2 POST TENSIONED OR STRESSED IN PLACE CONCRETE

	UNIT	MATERIAL
Concrete - 6" Slab and Columns	SqFt	4.05
Formwork (Assume Flying Forms)	SqFt	4.30
Post Tensioning and Reinforcing Steel	SqFt	2.80
TOTAL		11.15

.3 TILT UP CONSTRUCTION (Incl. Concrete Columns)

	UNIT	MATERIAL
Concrete - 6" Slab	SqFt	3.80
Formwork	SqFt	2.50
Reinforcing Steel	SqFt	2.40
Tilting Up and Bracing	SqFt	1.20
TOTAL		9.90

.4 PNEUMATICALLY PLACED CONCRETE (Gunite)

	UNIT	1"	2"	3"
Walls - 1" Layers with Mesh	SqFt	2.85	4.55	5.45
Columns - 1" Layers with Mesh	SqFt	5.40	5.95	6.90
Roofs	SqFt	3.65	4.25	4.75

0305.0 PRECAST PRESTRESSED CONCRETE (L&M)

	Spans to Approximate	UNIT	COST
.1 BEAMS AND COLUMNS			
Beams	20'	LnFt	84.00
	30'	LnFt	98.00
Columns - 12" x 12" x 12' with Plates		LnFt	105.00
16" x 16" x 12' with Plates		LnFt	130.00
.2 DECKS (Flat) - 24", 40", 48", 72" and 96" Wide - 50 lb. Roof Load			
4"	12'	SqFt	4.40
6"	18'	SqFt	4.90
8"	24'	SqFt	5.30
10"	30'	SqFt	5.80
12"	50'	SqFt	7.00
.3 DOUBLE T (Flat)			
8' x 20" - 30" Deep - 60 lb. Roof Load	40'	SqFt	5.70
	50'	SqFt	6.00
	60'	SqFt	7.00
	70'	SqFt	7.80
	80'	SqFt	8.40
SINGLE T			
8' x 36" - 48" Deep	60'	SqFt	9.10
	80'	SqFt	10.00
	100'	SqFt	11.20
	120'	SqFt	12.60
Add for Floor Decks (100# Load)		SqFt	10%
Add for Sloped Installation to Decks & T's		SqFt	10%
Add for 2" Topping for Floors - If Required		SqFt	2.90
Add per Mobilization		Each	2,600.00

		UNIT	COST
0305.0	**PRECAST PRESTRESSED CONCRETE, Cont'd...**		
.4	WALLS		
	Add to Flat Decks 0305.2 Above	SqFt	1.60
	Add to Double T 0305.3 Above	SqFt	1.75
	Add for Core Wall Type	SqFt	1.55
	Add for Insulated Type	SqFt	1.65
0306.0	**PRECAST CONCRETE (M)**		
.1	BEAMS AND COLUMNS- Same as 0305.1		
.2	DECKS- Same as 0305.2		
.3	PANELS OR FACING UNITS (L&M)		
	Insulated - 8' x 16' x 12"		
	Exposed Aggregate		
	Gravel	SqFt	12.50
	Quartz & Marble	SqFt	14.50
	Trowel or Brush		
	Grey Cement	SqFt	12.00
	White Cement	SqFt	13.25
	Deduct for Non-Insulated 8' x 16' x 8"	SqFt	1.15
	Add for Banded	SqFt	.42
.4	PLANK (S)		
	2" x 24" 6' Span	SqFt	4.00
	3" x 24" 8' Span	SqFt	4.50

			LABOR	MATERIAL
.5	SPECIALTIES (S)			
	Curbs 6" x 10" x 8'	LnFt	1.95	6.00
	8" x 10" x 8'	LnFt	2.35	6.50
	Copings 5" x 13"	LnFt	5.30	8.60
	Sills & Stools 5" x 6"	LnFt	4.25	10.80
	Splash Blocks 3" x 36" x 16"	Each	8.70	21.60
	5" x 36" x 16"	Each	10.90	22.60
	8" x 36" x 16"	Each	14.50	27.00
	Medians	LnFt	8.10	35.00

			SPANS TO APPROX.		COST
0307.0	**PRECAST CEMENTITIOUS PLANK (L&M)**				
.1	EXPANDED MINERALS (Includes Clips, Rods & Grout)				
	3" x 24"		6'-0"	SqFt	3.10
	4" x 24"		9'-6"	SqFt	3.35
.2	GYPSUM (Metal Edged)				
	2" x 15"		4'-0"	SqFt	3.20
	2" x 15"		7'-0"	SqFt	3.40
.3	WOOD FIBRE				
	2" x 32"		3'-0"	SqFt	2.95
	2 1/2" x 32"		3'-6"	SqFt	3.05
	3" x 32"		4'-0"	SqFt	3.25
0308.0	**POURED IN PLACE CEMENTITIOUS CONCRETE (L&M)**				
.1	EXPANDED MINERALS - 2"			SqFt	1.95
.2	EXPANDED MINERALS & ASPHALT - 2"			SqFt	2.15
.3	GYPCRETE - 1/4"			SqFt	1.00
	1"			SqFt	1.30

All examples below are representative samples and have great variables because of design efficiencies, size, spans, weather, etc. All figures include a contractor's fee of 10% for Footings and Slabs on Ground (low risk) and a 15% fee for Walls, Columns, Beams and Slabs (high risk). All totals include material prices as follows: Concrete $75.00 CuYd, Lumber $600.00 M.B.F. (3 uses), Plywood $1.10 SqFt, and Reinforcing Steel $650.00 Ton. Taxes and insurance on Labor added at 35%. Includes General Conditions at 10%. CuYd Costs are for information only--not quick estimating.

	Unit	Unit Cost With ReinSteel	Unit Cost Without ReinSteel	CuYd Cost With ReinSteel
FOOTINGS (Including Hand Excavation)				
Continuous 24" x 12" (3 # 4 Rods)	LnFt	18.66	16.96	252.00
36" x 12" (4 # 4 Rods)	LnFt	24.17	23.87	217.00
20" x 10" (2 # 5 Rods)	LnFt	15.90	15.17	302.00
16" x 8" (2 # 4 Rods)	LnFt	13.57	15.77	407.00
Pad 24" x 24" x 12" (4 # 5 E.W.)	Each	65.72	48.82	443.00
36" x 36" x 14" (6 # 5 E.W.)	Each	132.50	93.96	398.00
48" x 48" x 16" (8 # 5 E.W.)	Each	220.48	180.36	371.00
WALLS (#5 Rods 12" O.C. - 1 Face)				
8" Wall (# 5 12" O.C. E.W.)	SqFt	13.04	11.66	522.00
12" Wall (# 5 12" O.C. E.W.)	SqFt	14.26	12.85	385.00
16" Wall (# 5 12" O.C. E.W.)	SqFt	15.90	14.58	318.00
Add for Steel - 2 Faces - 12" Wall	SqFt	1.70	-	45.00
Add for Pilastered Wall - 24" O.C.	SqFt	.80	-	21.00
Add for Retaining or Battered Type	SqFt	2.39	-	64.00
Add for Curved Walls	SqFt	4.40	-	127.00
COLUMNS				
Sq Cornered 8" x 8" (4 # 8 Rods)	LnFt	31.80	22.68	1,908.00
12" x 12" (6 # 8 Rods)	LnFt	52.36	41.36	1,410.00
16" x 16" (6 # 10 Rods)	LnFt	72.08	53.24	1,081.00
20" x 20" (8 # 20 Rods)	LnFt	94.34	68.47	916.00
24" x 24" (10 # 11 Rods)	LnFt	123.49	82.94	830.00
Round 8" (4 # 8 Rods)	LnFt	22.79	15.44	1,643.00
12" (6 # 8 Rods)	LnFt	35.83	21.92	1,166.00
16" (6 # 10 Rods)	LnFt	48.55	30.67	922.00
20" (8 # 20 Rods)	LnFt	68.90	45.79	848.00
24" (10 # 11 Rods)	LnFt	100.70	63.18	827.00
BEAMS				
Spandrel 12" x 48" (33 # Rein. Steel)	LnFt	122.96	96.66	830.00
12" x 42" (26 # Rein. Steel)	LnFt	108.12	83.70	832.00
12" x 36" (21 # Rein. Steel)	LnFt	87.98	72.47	792.00
12" x 30" (15 # Rein. Steel)	LnFt	78.97	71.28	848.00
8" x 48" (26 # Rein. Steel)	LnFt	101.76	93.64	1,018.00
8" x 42" (21 # Rein. Steel)	LnFt	97.52	83.92	1,073.00
8" x 36" (16 # Rein. Steel)	LnFt	89.57	74.52	1,203.00
Interior 16" x 30" (24 # Rein. Steel)	LnFt	70.38	64.48	563.00
16" x 24" (20 # Rein. Steel)	LnFt	69.96	60.26	700.00
12" x 30" (17 # Rein. Steel)	LnFt	66.78	59.40	694.00
12" x 24" (14 # Rein. Steel)	LnFt	58.30	51.62	787.00
12" x 16" (12 # Rein. Steel)	LnFt	48.76	39.96	975.00
8" x 24" (13 # Rein. Steel)	LnFt	56.71	48.71	1,134.00
8" x 16" (10 # Rein. Steel)	LnFt	43.46	37.80	1,304.00

SLABS

	Unit	Unit Cost With ReinSteel	Unit Cost Without ReinSteel	CuYd Cost With ReinSteel
Solid -				
4" Thick	SqFt	8.37	6.70	678.40
5" Thick	SqFt	9.65	7.40	625.40
6" Thick	SqFt	10.60	7.13	572.40
7" Thick	SqFt	11.98	9.61	551.20
8" Thick	SqFt	13.46	11.45	546.96
Deduct for Post Tensioned Slabs	SqFt	.80	-	-
Pan -				
Joist -20" Pan - 10" x 2"	SqFt	11.66	10.15	-
12" x 2"	SqFt	13.14	11.12	-
30" Pan - 10" x 2 1/2"	SqFt	11.66	9.94	-
12" x 2 1/2"	SqFt	13.14	11.02	-
Dome -19" x 19" - 10" x 2"	SqFt	11.66	9.88	-
12" x 2"	SqFt	13.25	9.94	-
30" x 30" - 10" x 2 1/2"	SqFt	11.45	9.83	-
12" x 2 1/2"	SqFt	12.51	9.83	-

COMBINED COLUMNS, BEAMS AND SLABS

	Unit	Unit Cost With ReinSteel		
20' Span	SqFt	16.96	-	-
30' Span	SqFt	18.02	-	-
40' Span	SqFt	19.08	-	-
50' Span	SqFt	20.67	-	-
60' Span	SqFt	24.70	-	-

STAIRS (Including Landing)

	Unit			CuYd Cost With ReinSteel
4' Wide 10' Floor Heights 16 Risers	Riser	153.70	-	1,240.00
5' Wide 10' Floor Heights 16 Risers	Riser	185.50	-	1,325.00
6' Wide 10' Floor Heights 16 Risers	Riser	206.70	-	1,272.00

SLABS ON GROUND

	Unit	Unit Cost With Mesh	CuYd Cost With Mesh
4" Concrete Slab (6 6/10 - 10 Mesh - 5 1/2 Sack Concrete, Trowel Finished, Cured & Truck Chuted)	SqFt	3.15	255.00
Add per Inch of Concrete	SqFt	.53	
Add per Sack of Cement	SqFt	.13	
Deduct for Float Finish	SqFt	.08	
Deduct for Brush or Broom Finish	SqFt	.05	
Add for Runway and Buggied Concrete	SqFt	.16	
Add for Vapor Barrier (4 mil)	SqFt	.13	
Add for Sub - Floor Fill (4" sand/gravel)	SqFt	.39	
Add for Change to 6 6/8 - 8 Mesh	SqFt	.07	
Add for Change to 6 6/6 - 6 Mesh	SqFt	.16	
Add for Sloped Slab	SqFt	.20	
Add for Edge Strip (sidewalk area)	SqFt	.19	
Add for 1/2" Expansion Joint (20' O.C.)	SqFt	.07	
Add for Control Joints (keyed/ dep.)	SqFt	.15	
Add for Control Joints (joint filled)	SqFt	.16	
Add for Control Joints - Saw Cut (20' O.C.)	SqFt	.27	
Add for Floor Hardener (1 coat)	SqFt	.12	
Add for Exposed Aggregate - Washed Added	SqFt	.37	
Retarding Added	SqFt	.33	
Seeding Added	SqFt	.38	
Add for Light Weight Aggregates	SqFt	.40	
Add for Heavy Weight Aggregates	SqFt	.27	
Add for Winter Production Loss/Cost	SqFt	.32	

See 0206.2 for Curbs, Gutters, Walks and other Exterior Concrete
See 0302.53 for Topping and Leveling of Floors

	UNIT	COST
TOPPING SLABS		
2" Concrete	SqFt	2.40
3" Concrete	SqFt	2.90
No Mesh or Hoisting		
PADS & PLATFORMS (Including Form Work & Reinforcing)		
4"	SqFt	6.00
6"	SqFt	7.50
PRECAST CONCRETE ITEMS		
Curbs 6" x 10" x 8"	LnFt	10.00
Sills & Stools 6"	LnFt	22.00
Splash Blocks 3" x 16"	Each	54.00
MISCELLANEOUS ADDITIONS TO ABOVE CONCRETE IF NEEDED		
Abrasives - Carborundum - Grits	SqFt	.65
Strips	LnFt	2.40
Bushammer - Green Concrete	SqFt	1.55
Cured Concrete	SqFt	2.10
Champers - Plastic 3/4"	LnFt	.85
Wood 3/4"	LnFt	.47
Metal 3/4"	LnFt	.77
Colors Dust On	SqFt	.62
Integral (Top 1")	SqFt	1.15
Control Joints - Asphalt 1/2" x 4"	LnFt	.82
1/2" x 6"	LnFt	.93
PolyFoam 1/2" x 4"	LnFt	.88
Dovetail Slots 22 Ga	LnFt	.82
24 Ga	LnFt	1.05
Hardeners Acrylic and Urethane	SqFt	.20
Epoxy	LnFt	2.15
Epoxy	SqFt	.27
Rubber Asphalt	LnFt	.65
Moisture Proofing - Polyethylene 4 mil	SqFt	.16
6 mil	SqFt	.19
Non-Shrink Grouts - Non - Metallic	CuFt	34.00
Aluminum Oxide	CuFt	24.00
Iron Oxide	CuFt	37.00
Reglets Flashing	LnFt	1.35
Sand Blast - Light	SqFt	1.15
Heavy	SqFt	2.10
Shelf Angle Inserts 5/8"	LnFt	4.00
Stair Nosings Steel - Galvanized	LnFt	5.70
Tongue & Groove Joint Forms - Asphalt 3 1/2"	LnFt	1.65
Wood 3 1/2"	LnFt	1.15
Metal 3 1/2"	LnFt	1.55
Treads - Extruded Aluminum	LnFt	8.25
Cast Iron	LnFt	9.00
Water Stops Center Bulb - Rubber - 6"	LnFt	9.50
9"	LnFt	17.00
Polyethylene - 6"	LnFt	3.60
9"	LnFt	4.50

DIVISION #4 - MASONRY

Wage Rates (Including Fringes) & Location Modifiers
July 2001-2002

	Metropolitan Area	Bricklayer		Tender		Wage Rate Location Modifier
1.	Akron	*	34.16	*	27.09	115
2.	Albany-Schenectady-Troy		32.12		25.49	116
3.	Atlanta		20.72		13.68	74
4.	Austin	**	22.05	**	15.54	74
5.	Baltimore		24.94		17.85	90
6.	Birmingham	**	22.78	**	17.36	72
7.	Boston		41.56		30.53	145
8.	Buffalo-Niagara Falls		36.28		30.89	127
9.	Charlotte	**	19.77	**	13.61	66
10.	Chicago-Gary	*	37.99	*	32.96	126
11.	Cincinnati		28.31		23.96	94
12.	Cleveland		33.23		29.34	113
13.	Columbus	*	30.23	*	22.85	105
14.	Dallas-Fort Worth	**	20.23	**	14.14	69
15.	Dayton	*	30.37	*	23.52	101
16.	Denver-Boulder	*	24.91	*	15.96	83
17.	Detroit	*	38.30	*	29.04	126
18.	Flint	*	32.56	*	24.58	110
19.	Grand Rapids	*	32.69	*	22.36	110
20.	Greensboro-West Salem	**	19.92	**	13.63	66
21.	Hartford-New Britain	*	34.72	*	27.74	118
22.	Houston	**	22.63	**	16.49	82
23.	Indianapolis		29.79		23.37	101
24.	Jacksonville	**	23.90	**	15.63	75
25.	Kansas City		31.10		23.55	97
26.	Los Angeles-Long Beach	*	35.07	*	29.34	128
27.	Louisville	**	22.29	**	18.84	80
28.	Memphis	**	23.89	**	13.14	88
29.	Miami	**	23.91	**	16.46	80
30.	Milwaukee	*	33.89	*	28.72	115
31.	Minneapolis-St. Paul		34.46		29.59	112
32.	Nashville	**	22.69	**	15.27	78
33.	New Orleans	**	21.22	**	14.23	74
34.	New York	*	50.07	*	40.97	177
35.	Norfolk-Portsmouth	**	22.42	**	15.27	76
36.	Oklahoma City	**	23.78	**	16.17	88
37.	Omaha-Council Bluffs	**	24.65	**	19.53	82
38.	Orlando	**	24.89	**	18.58	84
39.	Philadelphia	*	37.65	*	31.57	127
40.	Phoenix	**	27.95	**	17.60	94
41.	Pittsburgh		32.98		24.00	111
42.	Portland	*	34.95	*	27.88	116
43.	Providence-Pawtucket	*	32.60	*	26.86	110
44.	Richmond	**	22.63	**	15.16	76
45.	Rochester	*	33.80	*	26.26	114
46.	Sacramento	*	41.61	*	30.07	138
47.	St. Louis		33.72		29.39	112
48.	Salt Lake City-Ogden	**	21.07	**	16.87	68
49.	San Antonio	**	20.49	**	12.42	67
50.	San Diego	*	31.90	*	29.92	127
51.	San Francisco-Oakland-San Jose		41.61		30.09	137
52.	Seattle-Everett		34.83		29.80	112
53.	Springfield-Holyoke-Chicopee		32.32		31.19	110
54.	Syracuse	*	29.97	*	24.14	99
55.	Tampa-St. Petersburg	**	24.90	**	17.31	85
56.	Toledo	*	31.71	*	25.33	107
57.	Tulsa	**	24.05	**	16.14	88
58.	Tucson	**	28.15	**	20.23	95
59.	Washington D.C.	*	25.67	*	17.13	87
60.	Youngstown-Warren	*	31.15	*	26.91	106
	AVERAGE		**29.40**		**22.56**	

Note: Labor Rates Increased by .15 for Tending (average)
        Impact Ratio: Labor 65%, Material 35%
*   Contract Not Settled - Wage Interpolated
**   Non Signatory or Open Shop Rate

			PAGE
0401.0	**BRICK MASONRY (04210)**		**4-4**
.1	FACE BRICK		4-4
.11	Conventional		4-4
.12	Econo		4-4
.13	Panel		4-4
.14	Norman		4-4
.15	King		4-4
.16	Norwegian		4-4
.17	Saxon - Jumbo		4-4
.18	Adobe - Mexican		4-4
.2	COATED(Ceramic Veneer) (04250)		4.4
.3	COMMON BRICK		4-4
.4	FIRE BRICK (04550)		4-4
0402.0	**CONCRETE BLOCK (04220)**		**4-5**
.1	CONVENTIONAL		4-5
.11	Foundation		4-5
.12	Backup		4-5
.13	Partitions		4-5
.2	ORNAMENTAL OR SPECIAL EFFECT		4-6
.21	Shadowall and Hi-Lite		4-6
.22	Scored		4-6
.23	Breakoff		4-6
.24	Split Face		4-6
.25	Rock Faced		4-6
.26	Adobe (Slump)		4-6
.3	SCREEN WALL		4-6
.4	INTERLOCKING		4-6
.5	SOUND		4-6
.6	BURNISHED		4-7
.7	PREFACED UNITS		4-7
0403.0	**CLAY BACKING AND PARTITION TILE (04240)**		**4-7**
0404.0	**CLAY FACING TILE (GLAZED) (04245)**		**4-8**
.1	6 T SERIES		4-8
.2	8 W SERIES		4-8
0405.0	**GLASS UNITS (04270)**		**4-8**
0406.0	**TERRA COTTA (04280)**		**4-8**
0407.0	**MISCELLANEOUS UNITS**		**4-8**
.1	COPING TILE		4-8
.2	FLUE LINING (04452)		4-8
0408.0	**NATURAL STONE (04400)**		**4-9**
.1	CUT STONE (04420)		4-9
.11	Limestone		4-9
.12	Marble		4-9
.13	Granite		4-9
.14	Slate		4-9
.2	ASHLAR STONE		4-9
.21	Limestone		4-9
.22	Marble		4-9
.23	Granite		4-9
.3	ROUGH STONE (04410)		4-9
.31	Flagstone and Rubblestone (04400)		4-9
.32	Field Stone or Boulders		4-9
.33	Light Weight Boulders (Igneous)		4-9
.34	Light Weight Limestone		4-9

		PAGE
0409.0	**PRECAST VENEERS AND SIMULATED MASONRY (04430)**	**4-9**
.1	ARCHITECTURAL PRECAST STONE (04435)	4-9
.2	PRECAST CONCRETE	4-9
.3	MOSAIC GRANITE PANELS	4-9
0410.0	**MORTARS (04100)**	**4-10**
.1	PORTLAND CEMENT AND LIME	4-10
.2	MASONRY CEMENT	4-10
.3	EPOXY CEMENT	4-10
.4	HIGH EARLY CEMENT	4-10
.5	FIRE RESISTANT	4-10
.6	ADMIXTURES	4-10
.7	COLORING	4-10
0411.0	**CORE FILLING FOR REINFORCED CONCRETE MASONRY (04290)**	**4-10**
0412.0	**CORE AND CAVITY FILLING FOR INSULATED MASONRY**	**4-11**
.1	LOOSE	4-11
.2	RIGID	4-11
0413.0	**ACCESSORIES AND SPECIALTIES (04150)**	**4-12**
.1	ANCHORS (04170)	4-12
.2	TIES	4-12
.3	REINFORCEMENT (04160)	4-12
.4	FIREPLACE AND CHIMNEY ACCESSORIES	4-12
.5	CONTROL AND EXPANSION JOINTS (04180)	4-13
.6	FLASHINGS	4-13
0414.0	**CLEANING AND POINTING (04510)**	**4-13**
.1	BRICK AND STONE	4-13
.2	BLOCK	4-13
.3	FACING AND TILE	4-13
0415.0	**MASONRY RESTORATION (04500)**	**4-13**
.1	RAKING FILLING AND TUCKPOINTING	4-13
.2	SANDBLASTING	4-13
.3	STEAM CLEANING	4-13
.4	HIGH PRESSURE WATER	4-13
.5	BRICK REPLACEMENT	4-13
.6	WATERPROOFING	4-13
0416.0	**INSTALLATION OF STEEL AND MISCELLANEOUS EMBEDDED ITEMS**	**4-13**
0417.0	**EQUIPMENT AND SCAFFOLD**	**4-13**
.1	EQUIPMENT	4-13
.2	SCAFFOLD	4-13
4A	**QUICK ESTIMATING SECTIONS**	**4A-14, 4A-15, 4A-16, 4A-17, 4A-18**
4B	**TABLES FOR:** Mortar Quantities	**4B-19, 4B-20**
	UNITS PER SQUARE FOOT	
	Placed Per Man Day	
	HAULING AND UNLOADING COSTS	
	WEIGHTS PER UNIT	

The unit costs in this Division are priced as being done by masonry or General Contractors. See Section 4A for Total Subcontracted Cost. In general, the pricing is based on Modular Units, and materials F.O.B. job site in truckload lots. The basic labor units do include mortar mixing, low scaffolding, tending, and labor fringe benefits. Cleaning, high scaffold, equipment reinforcements, taxes and insurance, etc. must be added to basic labor and material units. The basis for Labor Unit Costs can be found as follows: as to labor rate mean, Page 4-1, and as to number of units placed per day, Table 4B. Unit Costs are based on an average crew of 4 Bricklayers to 3 Tenders (including Mortar Mixer and Operators) for Brick and Stone Work, and 1 Bricklayer to 1 Tender for Block and Tile Work.

0401.0	BRICK MASONRY (Bricklayers)	Nominal Dimensions	UNIT	LABOR	MATERIAL (Truckload)
.1	FACE BRICK (See Table 4B for Mfg. Size)				
.11	Conventional	8" x 2-2/3" x 4"			
	Running Bond		M pcs	675.00	510.00
	Common Bond (6 Cs. Headers)		M pcs	685.00	510.00
	Stack Bond		M pcs	706.00	510.00
	Dutch & English Bond		M pcs	742.00	510.00
	Flemish		M pcs	742.00	510.00
.12	Economy	8" x 4" x 4"	M pcs	814.00	738.00
.13	Panel (Triple)	8" x 8" x 4"	M pcs	1,339.00	1,664.00
		8" x 16" x 4"	M pcs	2,215.00	3,328.00
.14	Norman	12" x 2-2/3" x 4"	M pcs	798.00	728.00
.15	King	7-5/8", 8-5/8" & 9-5/8" x 2-5/8" x 4"	M pcs	731.00	510.00
.16	Norwegian	12" x 3-1/5" x 4"	M pcs	865.00	853.00
.17	Saxon-Utility 3" wall	12"x 4"x 3" (thin)	M pcs	1,056.00	1,019.00
	4" wall	12"x 4"x 4"	M pcs	1,071.00	1,050.00
	6" wall	12"x 4"x 6"	M pcs	1,288.00	1,487.00
	8" wall	16"x 3-5/8"x 3-5/8"	M pcs	1,468.00	1,914.00
.18	Meridian		M pcs	659.00	530.00
.19	Adobe Brick - Mexican	8" x 2-3/8"x 4"	M pcs	803.00	822.00
		12"x 2-3/8"x 4"	M pcs	927.00	728.00
.2	COATED BRICK (Ceramic Veneer)		M pcs	757.00	1,144.00
.3	COMMON BRICK - Clay		M pcs	541.00	312.00
	Concrete		M pcs	541.00	281.00
.4	FIRE BRICK - Light Duty (domestic)		M pcs	680.00	1,144.00
	Medium Duty		M pcs	690.00	1,820.00
	Heavy Duty		M pcs	706.00	2,392.00
	Add for Fireplaces		M pcs	536.00	-
	Add: Less than Truckload Lot(14M pcs)				15%
	Add: Full Size Brick		M pcs	4%	8%
	Add: Ceramic Coating - to Common Brick		M pcs	15%	80%
	Deduct: 3" thick Brick (if available)		M pcs	4%	7%
	Add: Ea.Addl. 10' in Fl.Hgt.(or floor)		M pcs	3%	-
	Add: Hauling & Unloading if required		M pcs		See Table 4B
	Add: Breakage & Cutting Allowance		-	-	3%
	Add: Piers & Corbels		M pcs	15%	-
	Add: Sills & Soldiers		M pcs	20%	-
	Add: Floor Brick (over 2")		M pcs	10%	5%
	Add: Weave & Herringbone Pattern		M pcs	20%	7%
	Add: Stack bond		M pcs	8%	-
	Add: Circular or Radius Work		M pcs	20%	-
	Add: Rock Faced		M pcs	10%	-
	Add: Two-Face Joint Striking		M pcs	15%	-
	Add: Arches (including Formwork)		M pcs	75%	20%
	Add: Sawing (Special or Excessive)		M pcs	25%	5%
	Add: Winter Work (below 40 degrees)				
	Production Loss		M pcs	10%	-
	Encls - Wall Area - 1 Side & 3 Uses		Sq Ft	.19	.17
	2 Sides & 3 Uses		Sq Ft	.29	.30
	Heaters & Fuel - Wall Area		Sq Ft	.09	.20
	Heat Mortar		Cu Yd	9.89	5.00
	Deduct: Residential Work		-	10%	-

Crew - 4 Bricklayers to 3 Tenders (Incl. Mortar Mixers & Operators).
See Section 0905.7 for Thin Veneer Brick and Stone (under 2").
See Section 0206.2 for Paver Brick for Sitework.

0402.0   CONCRETE BLOCK (Bricklayers)

	UNIT	LABOR	MATERIAL StdWt	LtWt
.1 CONVENTIONAL				
.11 Foundation or Not Struck - Deduct .30 from Labor Units below.				
.12 Backup or Struck One Face - Deduct .15 from Labor Units below.				
.13 <u>Partition or Struck Two Faces Used for Labor Units below:</u>				
12" x 8" x 16" Plain	Ea	2.16	1.53	1.94
Double or Single Corner	Ea	2.19	1.63	2.04
Bull Nose (Sgl or Dbl)	Ea	2.27	1.79	2.19
Bond Beam	Ea	2.21	1.70	2.09
Header	Ea	2.22	1.70	2.09
12" x 8" x 8" Half Length Block	Ea	2.06	1.53	1.94
12" x 4" x 16" Half High Block	Ea	2.06	1.57	1.99
12" x 8" x 8" Lintel	Ea	2.27	1.69	2.09
12" x 16" x 8" Lintel	Ea	2.42	1.74	2.14
12" x 8" x 16" Control Joint	Ea	2.27	1.80	2.24
Half Control Joint	Ea	2.06	1.80	2.19
Fire Rated (4 Hr.)	Ea	2.18	1.80	2.19
L Corner	Ea	2.29	1.63	1.99
Solid or Cap	Ea	2.25	1.84	2.19
8" x 8" x 16" Plain	Ea	2.01	1.14	1.45
Double or Single Corner	Ea	2.06	1.24	1.55
Bullnose (Sgl or Dbl)	Ea	2.08	1.41	1.71
8" x 8" x 8" Half Bullnose	Ea	2.03	1.41	1.71
8" x 8" x 16" Bond Beam	Ea	2.08	1.41	1.60
Header	Ea	2.07	1.41	1.60
8" x 8" x 8" Half Length Block	Ea	1.98	1.14	1.35
8" x 8" x 16" Half High Block	Ea	1.98	1.14	1.51
8" x 8" x 8" Lintel	Ea	2.06	1.33	1.61
8" x 16" x 8" Lintel	Ea	2.27	1.33	1.66
8" x 8" x 16" Control Joint	Ea	2.06	1.43	1.61
8" x 8" x 8" Half Control Joint	Ea	1.98	1.43	1.61
8" x 8" x 16" Fire Rated (2 Hr.)	Ea	2.01	1.28	1.56
(4 Hr.)	Ea	2.01	1.43	1.71
Solid or Cap	Ea	2.06	1.33	1.58
6" x8" x 16" Plain	Ea	1.94	.92	1.24
Bull Nose (Sgl or Dbl)	Ea	1.96	1.24	1.50
6" x8" x 8" Half Bullnose	Ea	1.91	1.24	1.50
6" x8" x 16" Bond Beam or Lintel	Ea	1.98	1.12	1.40
6" x8" x 8" Half Length Block	Ea	1.91	.96	1.24
6" x4" x 16" Half High Block	Ea	1.91	1.00	1.31
6" x8" x 16" Fire Rated (2 Hr.)	Ea	1.93	1.07	1.33
L Corner	Ea	2.11	1.17	1.45
Solid or Cap	Ea	1.96	1.26	1.55
4" x 8" x 16" Plain	Ea	1.87	.75	1.08
Bull Nose	Ea	1.90	1.02	1.29
Bond Beam or Lintel	Ea	1.92	.92	1.20
4" x 8" x 8" Half Length Block	Ea	1.82	.77	1.15
4" x 8"x 16" Half High Block	Ea	1.82	.82	1.12
L Corner	Ea	2.06	.97	1.25
Solid or Cap	Ea	1.93	1.07	1.35

0402.0 CONCRETE BLOCK (Bricklayers)

	UNIT	LABOR	MATERIAL StdWt	LtWt

.1 CONVENTIONAL

.13 Partition or Struck Two Faces , Cont'd...

	UNIT	LABOR	StdWt	LtWt
16" x 8" x 16" Plain	Ea	2.52	1.89	2.45
16" x 8" x 8" Half Length	Ea	2.37	1.87	2.45
14" x 8" x 16" Plain	Ea	2.21	1.86	2.40
14" x 8" x 8" Half Length	Ea	2.06	1.84	2.40
14" x 8" x 16" Bond Beam	Ea	2.11	2.04	2.55
10" x 8" x 16" Plain	Ea	2.01	1.43	1.84
10" x 8" x 8" Half Length	Ea	1.96	1.43	1.84
10" x 8" x 16" Bond Beam	Ea	2.06	1.58	1.99
3" x 8" x 16" Plain	Ea	1.85	.77	.92
Solid	Ea	1.91	.80	1.19

Add or Deduct to all Blockwork Above for:

	UNIT	LABOR	StdWt	LtWt
Add: Full Size Block	Ea	.07	.08	-
Add: Breakage and Cutting	Ea	.09	3%	3%
Add: Jamb and Sash Block	Ea	.15	.10	.10
Add: Stack Bond Work	Ea	.15	-	-
Add: Radius or Circular Work	Ea	.88	-	-
Add: Pilaster, Pier or Pedestal Work	Ea	.41	-	-
Add: Knock-out Block	Ea	.41	.10	.10
Deduct: Light Weight Block-Installation	Ea	.10	-	-
Add: Winter Production Loss (below 40°)	Ea	10%	-	-
Add: Enclosures - Unit	Ea	.18	.08	.08
Add: Enclosures - Wall Area 1 Side	SqFt	.21	.16	.16
Add: Enclosures - Wall Area 2 Sides	SqFt	.31	.26	.26
Add: Heating & Fuel - Wall Area	SqFt	.09	.18	.20
Deduct: Residential Work	Ea	10%	-	-

See 0410 for Core Filling for Reinforced Masonry
See 0411 for Filling for Insulated Masonry

.2 ORNAMENTED OR SPECIAL EFFECT

	UNIT	LABOR	StdWt	LtWt
.21 Shadowall and Hi-Lite Block				
Add to Block Prices 0402.1	Ea	.13	.11	.13
.22 Scored Block				
Add to Block Prices 0402.1 - 1 Score	Ea	.19	.09	.09
- 2 Score	Ea	.24	.16	.16
.23 Break Off Block				
Add to Block Prices 0402.1 - Broken	Ea	.16	.66	.66
- Unbroken	Ea	.24	.61	.61
.24 Split Face Block				
Add to Block Prices 0402.1	Ea	.16	.24	.28
.25 Rock Faced Block				
Add to Block Prices 0402.1	Ea	.16	.17	.17
.26 Adobe (or Slump) Block				
Add to Block Prices 0402.1				
8" x 8" x 16"	Ea	.21	.31	-
4" x 8" x 16"	Ea	.19	.22	-
.3 SCREEN WALL - 4" x 12" x 12"	Ea	2.06	1.73	-
.4 INTERLOCKING				
.41 Epoxy Laid				
Add to Block Prices 0402.1	Ea	-	.12	.12
Deduct from Labor Costs 0402.1	Ea	.41	-	-
.42 Panelized (No Head Joint Mortar)				
Add to Block Prices 0402.1	Ea	-	.12	.12
Deduct from Labor Costs 0402.1	Ea	.46	-	-

0402.0 CONCRETE BLOCK, Cont'd...

		UNIT	LABOR	MATERIAL
.5	**SOUND BLOCK**			
	Add to Block Prices 0402.1	Ea	.26	.77
	Add for Fillers	Ea	.15	.26
.6	**BURNISHED BLOCK** - One Face Finish			
	(Truckload Price)			
	Plain 12" x 8" x 16"	Ea	2.47	3.57
	8" x 8" x 16"	Ea	2.27	2.96
	6" x 8" x 16"	Ea	2.16	2.65
	4" x 8" x 16"	Ea	2.06	2.35
	2" x 8" x 16"	Ea	2.01	2.24
	Shapes 12"	Ea	2.83	4.49
	8"	Ea	2.73	2.55
	6"	Ea	2.68	2.86
	4"	Ea	2.58	2.65
	2"	Ea	2.47	2.35
	Add for Bond Beam	Ea	.36	.87
	Add to above for Two Face Finish	Ea	.46	1.43
	Add to above for Scored Face	Ea	.31	.10
	Add for Color	Ea	-	.30
	Add for Bullnose	Ea	.36	.71
.7	**PREFACED UNITS** - Ceramic Glazed			
	(Truckload Price)			
	12" x 8" x 16" Stretcher	Ea	3.09	7.14
	Glazed 2 Face	Ea	3.61	11.22
	8" x 8" x 16" Stretcher	Ea	2.94	6.27
	Glazed 2 Face	Ea	3.24	10.51
	6" x 8" x 16" Stretcher	Ea	2.99	5.81
	Glazed 2 Face	Ea	3.19	10.61
	4" x 8" x 16" Stretcher	Ea	3.09	5.51
	Glazed 2 Face	Ea	3.09	9.89
	2" x 8" x 16" Stretcher	Ea	2.83	5.25
	4" x 16" x 16" Stretcher	Ea	5.25	18.36
	Add for Scored Block	Ea	.31	.41
	Add for Base Caps, Jambs, Headers	Ea	.57	1.94
	Add for Raked Joints and White Cement	Ea	.67	.12
	Add for Less than Truckload Lot	Ea	-	10%

Crew Ratio - 1 Block Layer to 1 Tender
(Tender includes Mixers & Operators)

Deduct 10% from Block Labor for Residential Construction

0403.0 CLAY BACKING AND PARTITION TILE
(Carload Price)

	UNIT	LABOR	MATERIAL
3" x 12" x 12"	Sq Ft	1.65	1.68
4" x 12" x 12"	Sq Ft	1.70	1.73
6" x 12" x 12"	Sq Ft	1.85	2.14
8" x 12" x 12"	Sq Ft	1.96	2.55

	UNIT	LABOR	MATERIAL
0404.0 CLAY FACING TILE			
.1 6T SERIES (SELECT QUALITY - CLEAR & FUNCTIONAL)			
2" x 5-1/3" x 12" Soap Stretcher (Open Back)	Ea	2.21	2.81
Soap Stretcher (Solid Back)	Ea	2.27	2.91
4" x 5-1/3" x 12" 1 Face Stretcher	Ea	2.37	4.00
Select 2 Face Stretcher	Ea	2.52	5.10
6" x 5-1/3" x 12" Select 1 Face Bond Stretcher	Ea	2.37	4.37
8" x 5-1/3" x 12" Select 1 Face Bond Stretcher	Ea	2.52	5.41
Shapes- Group 1 Square and B/N Corners	Ea	2.58	4.37
2 Sills and 2" Base	Ea	2.68	4.68
3 Sills and 4" Base	Ea	2.68	5.82
4 Starters and Octagons	Ea	2.94	10.92
5 Radials and End Closures	Ea	2.94	18.20
Add for Less than Carload Lots	Ea	-	10%
Add for Designer Colors	Ea	-	20%
Add for Base Only	Ea	-	40%
.2 8W SERIES (SELECT QUALITY - CLEAR & FUNCTIONAL)			
2" x 8" x 16" Soap Stretcher (Open Back)	Ea	2.99	4.78
Soap Stretcher (Solid Back)	Ea	3.04	5.30
4" x 8" x 16" 1 Face Stretcher	Ea	3.04	5.41
Select 2 Face Stretcher	Ea	3.24	8.32
6" x 8" x 16" Select 2 Face 6" Bond Stretcher	Ea	3.19	7.07
8" x 8" x 16" Select 1 Face 8" Bond Stretcher	Ea	3.30	8.63
Shapes- Group 1 Square and B/N Corners	Ea	3.19	7.49
2 Sills and 2" Base	Ea	3.71	9.67
3 Sills and 4" Base	Ea	3.81	12.06
4 Starters and Octagons	Ea	4.12	14.77
5 Radials and End Closures	Ea	4.22	36.40
Add for Less than Carload Lots	Ea	-	40%
Add for Designer Colors	Ea	-	20%
Add for Base Only	Ea	25%	-
0405.0 GLASS UNITS			
4" x 8" x 4"	Ea	3.09	4.89
6" x 6" x 4"	Ea	3.30	4.94
8" x 8" x 4"	Ea	3.66	6.24
12" x 12" x 4"	Ea	4.84	17.68
Add for Solar Reflective Type	Ea	-	40%
Accessories: Reinforcing & Expansion Joint	Ea	-	1.75
0406.0 TERRA-COTTA			
Unglazed	SqFt	2.16	5.20
Glazed	SqFt	2.47	6.34
Colored Glaze	SqFt	2.52	9.10
0407.0 MISCELLANEOUS UNITS			
.1 COPING TILE - 9" Double Slant	LnFt	3.40	8.32
12" Double Slant	LnFt	3.71	10.92
Bell	LnFt	3.71	9.15
Corners & Ends = 4 x Above Material Prices			
.2 FLUE LINING - 8" x 8"	LnFt	3.24	4.26
8" x 12"	LnFt	3.35	7.80
12" x 12"	LnFt	3.76	10.09
16" x 16"	LnFt	5.67	13.73
18" x 18"	LnFt	7.52	14.56
20" x 20"	LnFt	12.36	28.08
24" x 24"	LnFt	16.48	36.40

0408.0 NATURAL STONE (M) (Bricklayers)
(See 0906.0 for Stone under 2")

		SQUARE FOOT		CUBIC FOOT	
.1	CUT STONE	Labor	Matl.	Labor	Matl.
.11	Limestone				
	Indiana (Standard) and Alabama				
	Face Stone 3"	7.52	16.20	30.08	64.80
	4"	8.45	17.82	25.34	54.00
	Sills and Light Trim	-	-	32.96	64.80
	Copings and Heavy Sections	-	-	29.87	62.64
	Add for Select	2.37	-	12.36	
	Minnesota, Texas, Wisconsin, etc.				
	Face Stone 3"	7.52	21.60	30.08	86.40
	4"	8.55	23.76	25.96	71.28
	Sills and Light Trim	-	-	26.78	86.40
	Copings and Heavy Sections	-	-	25.75	84.24
.12	Marble				
	Face Stone 2"	7.80	27.00	32.96	113.40
	3" and 4"	8.50	29.16	29.87	91.80
	Sills and Light Trim	-	-	30.90	95.04
	Copings and Heavy Sections	-	-	28.84	81.00
.13	Granite				
	Face Stone 2"	8.03	23.76	42.23	135.00
	3"	9.58	25.92	38.11	102.60
	4"	11.12	29.16	33.99	89.64
	Sills and Light Trim	-	-	41.20	100.44
	Copings and Heavy Sections	-	-	38.11	95.04
.14	Slate 1 1/2"	8.55	25.38	51.50	140.40
.2	ASHLAR STONE (40 SqFt/Ton - 4" sawed bed)				
.21	Limestone				
	Indiana - Random	7.93	8.64		
	Coursed (2"-5"-8")	8.24	9.94		
	Minn., Ala., Texas, Wisc., etc.)				
	Split Face-1 Size 7 1/2"	8.45	9.72		
	Coursed (2"-5"-8")	9.79	10.48		
	Sawed/Planed Face-Random	8.14	10.58		
	Coursed (2"-5"-8")	7.83	11.66		
.22	Marble- Sawed Random	8.76	17.01		
	Coursed	8.96	17.71		
.23	Granite- Bushhammered-Random	9.37	19.44		
	Coursed	9.58	20.63		
.3	ROUGH STONE (30 s.f. per Ton)				
.31	Flagstone and Rubble 2"	8.76	7.78		
.32	Field Stone or Boulders	8.24	7.24		
.33	Lt.Wt. Boulders (Igneous) (260 s.f. Ton)				
	2" to 5" Veneer (Sawed Back)	6.95	6.80		
	3" to 10" Boulders	8.24	6.59		
.34	Lt.Wt. Limestone (90 s.f. Ton)				
	2" to 4"	7.93	8.37		

0409.0 PRECAST VENEERS AND SIMULATED MASONRY

.1	ARCHITECTURAL PRECAST STONE				
	Limestone and Gravel	7.31	17.28		
	Marble	7.93	24.30		
	Granite and Quartz	7.98	22.68		
.2	PRECAST CONCRETE	7.11	13.18		
.3	MOSAIC GRANITE PANELS	9.06	27.00		

0410.0 MORTARS

(See Table 4B for Quantities per Unit)
Material Cost used for Units Below (Truckload Prices)

	UNIT	LABOR	MATERIAL
Portland Cement - Gray	Bag- 94#	9.48	9.18
White	Bag- 94#	20.45	17.82
Masonry Cement	Bag- 70#	7.31	7.56
High Early Cement	Bag- 94#	-	9.72
Lime, Hydrated	Bag- 50#	5.67	5.94
Silica Sand	Bag-100#	-	6.48
Fire Clay	Bag-100#	-	19.44
Sand	CuYd	31.93	21.60
Add for Less than Truckload	-	-	20%
Labor Incl in Unit Costs:w/ Mortar Mixer	CuYd	22.66	-
Hand Mixed	CuYd	51.50	-

		UNIT	LABOR	MATERIAL
.1	PORTLAND CEMENT AND LIME MIX			
	(4.5 Bags Portland and 4.5 Bags Lime)			
	1:1:6 - Gray - Truckload	CuYd	23.69	81.00
	Less than Truckload	CuYd	23.69	95.04
	White	CuYd	23.69	108.00
	Add for Using Silica Sand	CuYd	20.60	32.40
.2	MASONRY CEMENT MIX (9 Bags Cement)			
	1:3 - Truckload	CuYd	23.69	66.96
	Less than Truckload	CuYd	23.69	77.76
.3	EPOXY CEMENT	CuYd	30.90	162.00
.4	FIRE RESISTANT CEMENT (Clay Mix 300# Mpc)	CuYd	30.90	91.00
.5	ADMIXTURES (Add to Mortar Prices above)			
	Waterproofing (1# sack cmt & $ .40/lb)	CuYd	16.48	8.10
	Accelerators (1 qt sack cmt & $1.00/lb)	CuYd	16.48	11.34
.6	COLORING (50# CuYd or 90# Mpc & $1.00/lb)	CuYd	18.54	43.20
	Standard Brick	M Pcs	22.66	93.96
	Add for Heating Mortar Materials	CuYd	10.30	6.48

10% allowed in above Prices for Shrinkage

0411.0 CORE FILLING FOR REINFORCED CONCRETE MASONRY

Based on 3500# Conc.@ $71 CuYd; Mortar @ $70 CuYd; and $23 CuYd to Mix
Deduct 50% from Labor Units if Truck Chuted or Pumped

	Void			Ready Mix		Job Mix	
	CuFt	%	Unit	L	M	L	M
12" x 8" x 16" Plain - 2 Cell	.43	48	Pc	.37	1.13	.70	1.35
	.48	53	SqFt	.39	1.24	.76	1.51
8" x 8" x 16" Plain - 2 Cell	.27	40	Pc	.28	.72	.48	.86
	.30	44	SqFt	.29	.79	.52	.95
6" x 8 " x16" Plain - 2 Cell	.19	40	Pc	.24	.52	.36	.62
	.21	44	SqFt	.25	.55	.39	.66
12" x 8" x 16" Bond Beam	.45	50	Pc	.38	1.19	.78	1.40
	.49	55	SqFt	.40	1.24	.84	1.53
8" x 8" x 16" Bond Beam	.21	32	Pc	.25	.56	.37	.68
	.23	35	SqFt	.26	.60	.39	.73
6" x 8" x 16" Bond Beam	.11	25	Pc	.19	.31	.27	.36
	.13	27	SqFt	.20	.33	.27	.41
12" x 16" x 8" Lintel	.52	58	pc	.45	1.35	.80	1.56
	.57	63	SqFt	.48	1.44	.87	1.70
8" x 16" x 8" Lintel	.27	40	Pc	.28	.72	.45	.83
	.30	44	SqFt	.30	.78	.49	.92
16" x 8" x 18" Pilaster Block	.96	72	Pc	.79	2.48	2.11	2.92
	1.08	81	LnFt	.88	2.81	2.42	3.35
Add for Vertical Rein.			Pc	.13	.14	.14	.14
(1 #4 Rods 16" O.C.)			SqFt	.14	.15	.15	.15

0412.0 CORE & CAVITY FILL FOR INSULATED MASONRY

.1	LOOSE	R Value-No Insulation	R Value- Insulated	UNIT	LABOR	MATERIAL
.11	Core Fill					
	12" Concrete Block	1.28	3.90	Pc	.31	.40
				SqFt	.34	.43
	Lt. Wt. Concrete Block	2.27	8.30	Pc	.31	.40
				SqFt	.34	.43
	10" Concrete Block	1.20	2.40	Pc	.27	.34
				SqFt	.29	.37
	Lt. Wt. Concrete Block	2.22	5.70	Pc	.26	.35
				SqFt	.29	.39
	8" Concrete Block	1.11	1.93	Pc	.21	.22
				SqFt	.22	.23
	Lt. Wt. Concrete Block	2.18	5.03	Pc	.21	.22
				SqFt	.22	.23
	6" Concrete Block	.90	1.79	Pc	.19	.15
				SqFt	.20	.16
	Lt. Wt. Concrete Block	1.65	2.99	Pc	.19	.15
				SqFt	.20	.16

		R Value	UNIT	LABOR	MATERIAL
	8" Brick	.88	SqFt	.20	.11
	6" Brick	.66	SqFt	.16	.08
	4" Brick	.44	SqFt	.14	.06
	Add to above for Expanded Mica	-	-	10%	200%
	Add to above for Perlite	-	-	-	100%
	10% added for spillage				

.12	Cavity Fill - Per Inch Wall				
	Expanded Styrene	3.8	SqFt	.20	.19
	Fiber Glass	2.2	SqFt	.21	.22
	Rock Wool	2.9	SqFt	.21	.22
	Cellulose	3.7	SqFt	.21	.20
	Add per added inch	-	-	40%	100%
	10% added for spillage				

.2	RIGID				
	Fiber Glass				
	1" - 3# Density	4.35	SqFt	.28	.32
	1.5"	6.52	SqFt	.31	.48
	2"	8.70	SqFt	.34	.64
	Expanded Styrene - Molded (White)				
	1" - 1# Density	4.33	SqFt	.28	.19
	1.5"	6.52	SqFt	.31	.28
	2"	8.70	SqFt	.34	.37
	Add for Embedded Nailer Type	-	-	-	.10
	Expanded Styrene - Extruded (Blue)				
	1" - 2# Density	5.40	SqFt	.28	.36
	1.5"	8.10	SqFt	.31	.48
	2"	10.80	SqFt	.34	.62
	Expanded Urethane				
	1"	7.14	SqFt	.30	.80
	2"	14.28	SqFt	.34	1.55
	Perlite				
	1"	2.78	SqFt	.27	.40
	2"	5.56	SqFt	.33	.80
	Add for Glued Applications	-	SqFt	.05	.02

0413.0 ACCESSORIES & SPECIALTIES - MATERIAL

		UNIT	LABOR	MATERIAL
.1	**ANCHORS**			
	Dovetail-Brick 12 ga Galvanized x 3 - 1/2"	Ea	.16	.23
	Brick 16 ga Galvanized x 3 - 1/2"	Ea	.16	.20
	x 5"	Ea	.16	.28
	Stone - Stainless	Ea	.16	.95
	Rigid Wall - 1 - 1/4"x 3/6"x 8" - Galv	Ea	.16	.90
	Stic - Kips and Clamp	Ea	.20	.20
	Strap Z - 1/8"x 1"x 6" or 8" - Galv	Ea	.34	1.05
	1/8" x 1"x 6" - SS	Ea	.34	2.30
	3/16" x 1 - 1/4" x 8" - Galv	Ea	.34	1.55
	Split Tail - 3/16" x 1 - 1/4" x 6" - Galv	Ea	.34	1.50
	1/8" x 1 - 1/2"x 6" - SS	Ea	.34	3.10
	Wall Clips - 2"x 3"	Ea	.34	.18
	Wall Plugs - Wood Filled	Ea	.34	.20
.2	**TIES - WIRE**			
	Galv Z - 3/16" Wire - 6" & 8"	Ea	.16	.16
	Zinc Z - 3/16" Wire - 6" & 8"	Ea	.16	.36
	Copperweld Z - 3/16" Wire - 6"	Ea	.16	.42
	3/16" Wire - 8"	Ea	.16	.50
	Adjustable Rect. - 3 /16" Wire 3 - 1/4"	Ea	.16	.28
	3/16" Wire 4 - 3/4"	Ea	.16	.32
	Corrugated-Strap - 26 ga - 7/8" x 7"	Ea	.16	.04
	22 ga - 7/8" x 7"	Ea	.16	.05
	16 ga - 7/8" x 7"	Ea	.16	.08
	Hardware Cloth - 2" x 5" x 1/4" - 23 ga	Ea	.16	.13
	23 ga x 1/4" x 30"	SqFt	.22	.58
.3	**REINFORCEMENT**			
	#9 Welded Wire - Ladder Type - 4"	LnFt	.13	.08
	6"	LnFt	.13	.09
	8"	LnFt	.14	.11
	10"	LnFt	.14	.12
	12"	LnFt	.15	.13
	16"	LnFt	.17	.15
	Add for 3/16" Wire H.D.	LnFt	-	30%
	Add for Truss Type	LnFt	-	20%
	Add for Rect. Ties - Composite & Cavity	LnFt	.03	.05
	Add for Rect. Adj. Ties - Comp & Cavity	LnFt	.03	.08
	Reinforcement Rods - Horizontal Placement			
	3/8" or #3 .376 lb	LnFt	15	.17
	1/2" or #4 .668 lb	LnFt	.19	.26
	5/8" or #5 1.043 lb	LnFt	.28	.40
	Add for Vertical Placement	LnFt	25%	-
.4	**FIREPLACE AND CHIMNEY ACCESSORIES**			
	Ash Dumps and Thimbles - 6" x 8"	Ea	25.00	10.00
	Clean Cut Doors, Cast Iron - 8" x 8"	Ea	22.00	18.00
	12" x 12"	Ea	29.00	40.00
	Dome Dampers, Cast Iron - 24" x 16"	Ea	20.00	46.00
	30" x 16"	Ea	24.00	60.00
	36" x 20"	Ea	35.00	87.00
	42" x 20"	Ea	48.00	92.00
	48" x 24"	Ea	65.00	125.00

0413.0 ACCESSORIES & SPECIALTIES, Cont'd...

		UNIT	LABOR	MATERIAL 1/2"	1"
.5	CONTROL AND EXPANSION JOINT MATERIAL				
	Asphalt - Fibre	SqFt	.52	.95	1.79
	Polyethylene Foam	SqFt	.52	1.05	1.89
	Sponge Rubber	SqFt	.52	5.99	7.35
	Paper Fibre	SqFt	.52	.79	1.47
	Also in Thk. of 1/8", 1/4", 3/8" & 3/4"				

		UNIT	LABOR	2 oz	3 oz
.6	FLASHING, THRU WALL				
	Copper Between Polyethylene	SqFt	.67	1.16	1.42
	Rein Paper - 2 oz/1.05	SqFt	.67	1.31	1.94
	Asphalted Fabric - 3 oz	SqFt	.67	1.73	2.31
	Poly Vinyl Chloride - 20 mil	SqFt	.52	-	.21
	30 mil	SqFt	.57	-	.37
	Add for Adhesives			.05	.05

0414.0 CLEANING & POINTING (See 0414 for Restoration)

		UNIT	LABOR		
.1	BRICK & STONE- Acid & Other Chemicals	SqFt	.39	-	.02
	Soap & Water	SqFt	.34	-	-
.2	BLOCK - 1 Face	SqFt	.19	-	-
.3	FACING TILE AND BLOCK				
	Point - White Cement and Sand	SqFt	.67	-	.07
	Silica Sand	SqFt	.82	-	.12
	Add for Scaffold (See 0417)				

0415.0 MASONRY RESTORATION (L&M) (Bricklayers)

		UNIT	Cost
.1	RAKING, FILLING AND TUCKPOINTING		
	Limited - Standard Brick & 3/8" Joint	SqFt	2.68
	Heavy	SqFt	3.31
	All	SqFt	6.62
	Deduct for Stone	SqFt	20%
	Deduct for 4" x 12" Brick	SqFt	30%
	Add for Larger & Smaller Joints	SqFt	25%
.2	SAND BLAST		
.22	Brick -Wet	SqFt	1.47
	Dry	SqFt	1.37
.23	Block -Wet	SqFt	1.37
	Dry	SqFt	1.26
.3	STEAM CLEAN AND CHEMICAL CLEANING	SqFt	1.21
.4	HIGH PRESSURE WATER CLEANING (Hydro)	SqFt	.89
.5	BRICK REPLACEMENT -Standard	Each	23.10
	4" x 12"	Each	18.90
	Add for Scaffolding	SqFt	.63
	Add for Protection	SqFt	.26
.6	WATER PROOFING - CLEAR	SqFt	.76
	See 0702.3 for Waterproofing Masonry		
	See 0711.0 for Caulking		

0416.0 INSTALLATION OF STEEL & MISC. EMBEDDED ITEMS
See Division 0502

0417.0 EQUIPMENT AND SCAFFOLD

		UNIT	LABOR	MATERIAL
.1	EQUIPMENT AVERAGE OF LABOR			
	See 1-7A for Rental Rates			
.2	SCAFFOLD - INCLUDING PLANK			
	Tubular Frame- to 40' - Exterior	SqFt	.35	.19
	to 16' - Interior	SqFt	.32	.19
	Swing State (Incl. Outriggers & Anchorage)	SqFt	.31	.21

The costs below are average but can vary greatly with Quantity, Weather, Area Practices and many other factors. They are priced as total contractor's price with an overhead and fee of 10% included. Units include mortar, ties, cleaning, and reinforcing. Not included are scaffold, hoisting, lintels, accessories, heat, and enclosures. Also added are 27% Insurance on Labor; 10% for General Conditions, Equipment and Tools; and 5% on Material for Sales Taxes. Crew Size is based on 4 Bricklayers to 3 Tenders for Brick and Stone Work, and 4 Bricklayers to 4 Tenders for Block and Tile. See Table 3 for Production Standards. Deduct 5% for Residential Work.

Example:	.1	Face Brick 8" x 2 2/3" x 4"	UNIT	LABOR	MATERIAL	TOTAL
	.11	Running Bond	Each	.65	.49	1.14
		Mortar and Clean Brick	Each	.10	.09	.19
				.75	.58	1.33
		Add: 27% Labor Burden, 5% Sales Tax		.20	.03	.23
				.95	.61	1.56
		Add: 10% General Conditions & Equipment				.16
						1.72
		Add: 10% Overhead and Fee				.17
		Total Brick Cost	Each Unit			1.89
			or SqFt			12.82

0401.0 BRICK MASONRY

			COST SqFt	COST Unit
.1	FACE BRICK			
.11	Conventional (Modular Size)			
	Running Bond - 8" x 2 2/3" x 4"		12.97	2.00
	Common Bond - 6 Course Header		15.75	2.01
	Stack Bond		13.97	2.07
	Dutch & English Bond - Every Other Course Header		21.58	2.13
	Every Course Header		23.94	2.13
	Flemish Bond - Every Other Course Header		16.17	2.13
	Every Course Header		19.11	2.13
	Add: If Scaffold Needed		.74	.11
	Add: For each 10' of Floor Hgt. or Floor (3%)		.35	.05
	Add: Piers and Corbels (15% to Labor)		1.05	.16
	Add: Sills and Soldiers (20% to Labor)		1.40	.21
	Add: Floor Brick (10% to Labor)		.74	.11
	Add: Weave & Herringbone Patterns (20% to Labor)		1.40	.21
	Add: Stack Bond (8% to Labor)		.57	.08
	Add: Circular or Radius Work 20% to Labor)		1.40	.21
	Add: Rock Faced & Slurried Face (10% to Labor)		.74	.11
	Add: Arches (75% to Labor)		4.62	.68
	Add: For Winter Work (below 40°)			
	Production Loss (10% to Labor)		.74	.11
	Enclosures - Wall Area Conventional		.74	.11
	Heat and Fuel - Wall Area Conventional		.24	.04
.12	Econo - 8" x 4" x 3"		11.34	2.52
.13	Panel - 8" x 8" x 4"		9.92	4.41
.14	Norman - 12" x 2 2/3" x 4"		11.06	2.46
.15	King Size - 10" x 2 5/8" x 4"		9.87	2.10
.16	Norwegian - 12" x 3 1/5" x 4"		9.89	2.64
.17	Saxon-Utility - 12" x 4" x 3"		9.45	3.15
	12" x 4" x 4"		9.61	3.20
	12" x 4" x 6"		11.97	3.99
	12" x 4" x 8"		14.81	4.94
.18	Adobe - 12" x 3" x 4"		10.33	2.58
.2	COATED BRICK (Ceramic)		18.90	2.81
.3	COMMON BRICK (Clay, Concrete and Sand Lime)		9.61	1.42
.4	FIRE BRICK -Light Duty		17.75	2.63
	Heavy Duty		20.69	3.08
	Deduct for Residential Work - All Above		-	5%

Example:		UNIT	LABOR	MATERIAL	TOTAL
12" x 8" x 16" Concrete Block - Partition		Each	2.10	1.50	3.60
Mortar (Labor in Block Placing Above)		Each	-	.20	.20
Reinforcing Wire 16" O.C. Horizontal		Each	.08	.07	.15
Clean Block - 2 Faces		Each	.25	-	.25
			2.43	1.77	4.20
Add: 27% Taxes and Insurance on Labor					.65
Add: 5% Sales Tax on Material					.09
					4.94
Add: 10% General Cond., Equip. and Tools					.49
					5.43
Add: 10% Overhead and Profit					.54
Total Block Cost		Each			5.97
		or SqFt			6.57

Not Included - Scaffold, Hoisting Equipment, Lintels, Embedded Accessories, Insulation, Heat and Enclosures. (See "Adds" Below.)

0402.0 CONCRETE BLOCK

		Std Wt Cost		Lt Wt Cost	
.1	CONVENTIONAL (Struck 2 Sides - Partitions)	SqFt	Unit	SqFt	Unit
	12" x 8" x 16" Plain	6.77	6.15	7.29	6.49
	Bond Beam (w/ Fill-Reinforcing)	9.37	8.34	9.96	8.86
	12" x 8" x 8" Half Block	6.72	5.97	7.05	6.26
	Double End - Header	7.05	6.26	7.73	6.87
	8" x 8" x 16" Plain	5.95	5.29	6.23	5.56
	Bond Beam (w/ Fill-Reinforcing)	7.18	6.39	7.42	6.59
	8" x 8" x 8" Half Block	5.81	5.17	5.85	5.20
	Double End - Header	6.37	5.67	7.42	6.59
	6 "x 8" x 16" Plain	5.25	4.67	5.97	5.33
	Bond Beam (w/ Fill-Reinforcing)	6.28	5.58	6.13	5.48
	Half Block	5.23	4.66	5.94	5.31
	4" x 8" x 16" Plain	5.05	4.49	5.05	4.49
	Half Block	4.92	4.38	4.82	4.28
	16" x 8" x 16" Plain	8.03	7.15	8.81	7.83
	Half Block	7.64	6.80	8.45	7.52
	14" x 8" x 16" Plain	7.42	6.59	8.11	7.21
	Half Block	6.95	6.18	7.00	6.90
	10" x 8" x 16" Plain	6.13	5.46	8.14	6.23
	Bond Beam (w/ Fill-Reinforcing)	6.95	6.18	6.64	7.31
	Half Block	6.08	5.41	0.22	5.92
	Deduct: Block Struck or Cleaned One Side	.22	.20	.22	.20
	Deduct: Block Not Struck or Cleaned Two Sides	.44	.39	.44	.39
	Deduct: Clean One Side Only	.18	.15	.18	.15
	Deduct: Lt. Wt. Block (Labor Only)	.14	.13	.14	.13
	Add: Jamb and Sash Block	.39	.35	.39	.35
	Add: Bullnose Block	.58	.52	.58	.52
	Add: Pilaster, Pier, Pedestal Work	.64	.58	.64	.58
	Add: Stack Bond Work	.25	.22	.25	.22
	Add: Radius or Circular Work	1.14	1.05	1.14	1.05
	Add: If Scaffold Needed	.67	.60	.67	.60
	Add: Winter Production Cost (10% Labor)	.37	.33	.37	.33
	Add: Winter Enclosing - Wall Area	.41	.37	.41	.37
	Add: Winter Heating - Wall Area	.31	.28	.31	.28
	Add: Core & Beam Filling - See 0411 (Page 4A-18)				
	Add: Insulation - See 0412 (Page 4A-18)				

Deduct for Residential Work - 5%

		Std Wt Cost		Lt Wt Cost	
0402.0	**CONCRETE BLOCK, Cont'd...**	**SqFt**	**Unit**	**SqFt**	**Unit**
.2	ORNAMENTAL OR SPECIAL EFFECT -				
	Add to Units Above in 0402.1				
	Shadow Wall	.33	.30	.35	.32
	Scored Block	.40	.35	.38	.35
	Breakoff	1.23	1.11	1.23	1.11
	Split Face	.50	.45	.66	.59
	Rock Faced	.52	.44	.47	.47
	Adobe (Slump)	.71	.63	-	-
.3	SCREEN WALL - 4" x 12" x 12"	4.78	4.78	-	-
.4	INTERLOCKING - Deduct from Units 0402.1				
	Epoxy Laid	.44	.40	.42	.40
	Panelized	.64	.58	.62	.58
.5	SOUND BLOCK - Add to Block Prices 0402.1			1.51	1.35
.6	BURNISHED - 12" x 8" x 16"			10.30	9.36
	8" x 8" x 16"			7.80	7.07
	6" x 8" x 16"			7.33	6.66
	4" x 8" x 16"			6.66	6.03
	2" x 8" x 16"			6.29	5.72
	Add for Shapes			1.20	1.04
	Add for 2 Faced Finish			2.86	2.60
	Add for Scored Finish - 1 Face			.62	.57
.7	PREFACED UNITS -12" x 8" x 16" Stretcher			15.18	13.52
	(Ceramic Glazed) 12" x 8" x 16" Glazed 2 Face			21.32	18.93
	8" x 8" x 16" Stretcher			14.04	12.48
	Glazed 2 Face			19.24	17.16
	6" x 8" x 16" Stretcher			13.00	11.54
	Glazed 2 Face			17.89	15.91
	4" x 8" x 16" Stretcher			12.27	10.92
	Glazed 2 Face			18.51	16.43
	2" x 8" x 16" Stretcher			11.75	10.50
	4" x 16" x 16" Stretcher			35.05	31.20
	Add for Base, Caps, Jambs, Headers, Lintels			3.43	3.12
	Add for Scored Block			1.04	.94

		COST	
0403.0	**CLAY BACKING AND PARTITION TILE**	**SqFt**	**Unit**
	3" x 12" x 12"	4.63	4.63
	4" x 12" x 12"	4.84	4.84
	6" x 12" x 12"	5.77	5.77
	8" x 12" x 12"	6.14	6.14
0404.0	**CLAY FACING TILE (GLAZED)**		
.1	6T or 5" x 12" - SERIES		
	2" x 5 1/3" x 12" Soap Stretcher (Solid Back)	14.98	6.66
	4" x 5 1/2" x 12" 1 Face Stretcher	17.32	7.70
	2 Face Stretcher	20.80	9.26
	6" x 5 1/3" x 12" 1 Face Stretcher	21.06	9.36
	8" x 5 1/3" x 12" 1 Face Stretcher	26.42	11.75
.2	8W or 8" x 16" - SERIES		
	2" x 8" x 16" Soap Stretcher (Solid Back)	12.79	11.34
	4" x 8" x 16" 1 Face Stretcher	13.57	12.01
	2 Face Stretcher	16.28	14.40
	6" x 8" x 16" 1 Face Stretcher	15.91	14.14
	8" x 8" x 16" 1 Face Stretcher	17.68	15.60
	Add for Shapes (Average)	50%	50%
	Add for Designer Colors	-	20%
	Add for Less than Truckload Lots	-	10%
	Add for Base Only	-	25%

		COST	

0405.0	GLASS UNITS	SqFt	Unit
	4" x 8" x 4"	43.94	9.77
	6" x 6" x 4"	40.74	10.19
	8" x 8" x 4"	25.99	11.55
	12" x 12" x 4"	25.20	25.20
0406.0	TERRA-COTTA		
	Unglazed	9.45	9.45
	Glazed	13.13	13.13
	Colored Glazed	15.33	15.33

0407.0	FLUE LINING		Ln Ft
	8" x 12"		14.70
	12" x 12"		17.33
	16" x 16"		27.30
	18" x 18"		30.45
	20" x 20"		53.55
	24" x 24"		76.65

0408.0	NATURAL STONE			Sq Ft	Cu Ft
.1	CUT STONE				
		Limestone- Indiana and Alabama - 3"		31.19	123.90
			4"	34.65	103.95
		Minnesota, Wisc, Texas, etc.- 3"		36.75	147.00
			4"	41.48	123.90
.12		Marble - 2"		45.15	270.90
			3"	46.73	186.90
.13		Granite - 2"		40.95	245.70
			3"	44.63	178.50
.14		Slate - 1 1/2"		40.95	245.70
.2	ASHLAR- 4" Sawed Bed				
.21		Limestone- Indiana - Random		23.10	69.30
			Coursed - 2" - 5" - 8"	24.15	72.45
		Minnesota, Alabama, Wisconsin, etc.			
		Split Face - Random		24.89	74.55
			Coursed	26.46	78.75
		Sawed or Planed Face - Random		26.25	78.75
			Coursed	27.30	81.90
.22		Marble - Sawed - Random		34.65	103.95
			Coursed	35.70	107.10
.23		Granite - Bushhammered - Random		36.75	110.25
			Coursed	38.85	116.55
.24		Quartzite		26.25	78.75
.3	ROUGH STONE				
.31		Rubble and Flagstone		24.15	63.00
.32		Field Stone or Boulders		23.10	59.85
.33		Light Weight Boulders (Igneous)			
		2" to 4" Veneer - Sawed Back		19.95	-
		3" to 10" Boulders		21.00	-
.34		Light Weight Limestone		22.05	-

0409.0	PRECAST VENEERS AND SIMULATED MASONRY	Unit	Cost
.1	ARCHITECTURAL PRECAST STONE- Limestone	SqFt	30.98
	Marble	SqFt	36.75
.2	PRECAST CONCRETE	SqFt	24.36
.3	MOSAIC GRANITE PANELS	SqFt	35.18
	See 0904 for Veneer Stones (under 2")		

0410.0	MORTAR		
	Portland Cement and Lime Mortar - Labor in Unit Costs	CuYd	110.25
	Masonry Cement Mortar - Labor in Unit Costs	CuYd	99.75
	Mortar for Standard Brick - Material Only	SqFt	.40
	Add for less than Truckload	CuYd	12.60

		COST	
		SqFt	Unit
0411.0	**CORE FILLING FOR REINFORCED CONCRETE BLOCK**		
	Add to Block Prices in 0402.0 (Job Mixed)		
	12" x 8" x 16" Plain (includes #4 Rod 16" O.C. Vertical)	2.62	2.46
	Bond Beam (includes 2 #4 Rods)	2.72	2.55
	8" x 8" x 6" Plain (includes #4 Rod 16" O.C. Vertical)	1.81	1.72
	Bond Beam (includes 2 #4 Rods)	1.40	1.36
	6" x 8" x 16" Plain	1.24	1.19
	Bond Beam (includes 1 #4 Rod)	.91	.91
	10" x 8" x 16" Bond Beam (includes 2 #4 Rods)	2.43	2.32
	14" x 8" x 16" Bond Beam (includes 2 #4 Rods)	3.09	2.97
0412.0	**CORE AND CAVITY FILL FOR INSULATED MASONRY**		
	See Page 4-11 for R Values		
.1	LOOSE		
	Core Fill - Expanded Styrene		
	12" x 8" x 16" Concrete Block	.96	.94
	10" x 8" x 16" Concrete Block	.66	.67
	8" x 8" x 16" Concrete Block	.53	.55
	6" x 8" x 16" Concrete Block	.41	.44
	8"Brick - Jumbo - Thru the Wall	.37	.41
	6"Brick - Jumbo - Thru the Wall	.28	.32
	4"Brick - Jumbo	.21	.25
	Cavity Fill - per Inch		
	Expanded Styrene	.48	-
	Mica	.81	-
	Fiber Glass	.50	-
	Rock Wool	.50	-
	Cellulose	.47	-
.2	RIGID - Fiber Glass - 1" 3# Density	.75	-
	2"	1.20	-
	Expanded Styrene - Molded - 1" 1# Density	.54	-
	2"	.79	-
	Extruded - 1" 2# Density	.80	-
	2"	1.27	-
	Expanded Urethane - 1"	1.05	-
	2"	1.78	-
	Perlite - 1"	.79	-
	2"	1.21	-
	Add for Embedded Water Type	.14	-
	Add for Glued Applications	.06	-
0413.0	**ACCESSORIES AND SPECIALTIES**		
	See Page 4-12		
0414.0	**CLEANING AND POINTING**		
	Brick and Stone - Acid or Chemicals	.64	-
	Soap and Water	.58	-
	Block & Facing Tile - 1 Face (Incl. Hollow Metal Frames)	.32	-
	Point with White Cement	1.32	-
0415.0	**MASONRY RESTORATION** - See 0415.0		
0416.0	**INSTALLATION OF STEEL** - See 0502.0		
0417.0	**SCAFFOLD**		
.1	EQUIPMENT, TOOLS AND BLADES		
	Percentage of Labor as an average	-	7%
	See 1 - 7A for Rental Rates & New Costs		
.2	SCAFFOLD- Tubular Frame - to 40 feet - Exterior	.73	-
	Tubular Frame - to 16 feet - Interior	.68	-
	Swing Stage - 40 feet and up	.68	-

TABLE B

Modular Manufactured			Mortar CuFt	Unit Per	Units Placed	Haul & Unload	Unit
Size - Inches			M	SqFt	ManDay	M PCs	Wt.
l	h	d					

0401.0 BRICK MASONRY

.1 Face Brick

.11 Conventional

Running Bond	7-5/8 x 2-1/4 x 3-5/8		13	6.75	600	16.00	4.5
Common Bond							
Header Every 6th Course				ƒ1/6	570	-	-
Header Every 7th Course				ƒ1/7	570	-	-
Dutch and English Bond							
Header Every Other Course				ƒ1/2	510	-	-
Header Every Course				ƒ2/3	510	-	-
Flemish Bond							
Header Every Other Course				ƒ1/8	500	-	-
Header Every Course				ƒ1/3	500	-	-

.12	Econo	7-5/8 x 3-5/8 x 3-5/8			4.50	430	20.00	6.0
.13	Panel	7-5/8 x 7-5/8 x 3-5/8			2.25	280	40.00	13.5
		7-5/8 x 15-5/8 x 3-5/8			1.25	180	80.00	27.0
.14	Norman	11-5/8 x 2-1/4 x 3-5/8		16	4.50	450	20.00	6.0
.15	King	9-5/8 x 3-1/2 x 3-5/8		16	4.70	440	22.00	4.5
.16	Norwegian	11-5/8 x 2-13/16 x 3-5/8		16	3.75	370	22.00	8.0
.17	Saxon 4"	11-5/8 x 3-5/8 x 3-5/8		19	3.00	320	26.00	9.0
	6"	11-5/8 x 3-5/8 x 5-5/8		28	3.00	260	35.00	13.5
	8"	11-5/8 x 3-5/8 x 7-5/8		38	3.00	200	50.00	18.0

Deduct from above dimensions for 1/2" Joint 1/8" x 1/8" x 1/8".
Add to above dimensions for full size 3/8" x 0" x 2/8".

.2	Coated Brick (Glazed)			12	6.75	510	17.00	4.5
.3	Common Brick							
	Clay	7-14 x 2-1/16 x 3-1/2		13	6.75	660	15.00	4.5
	Concrete			13	6.75	660	15.00	4.5
.4	Fire Brick	8-0 x 2-1/4 x 3-1/2		13	6.65	660	15.00	7.0

0402.0 STANDARD AND LIGHT-WEIGHT CONCRETE BLOCK

12" Plain	11-5/8 x 7-5/8 x 15-5/8	65	1.125	190	110.00	5.0
1/2 Height	11-5/8 x 3-5/8 x 15-5/8		2.25		Lt Wt	38.0
1/2 Length	11-5/8 x 7-5/8 x 7-5/8		2.25			
8" Plain	7-5/8 x 7-5/8 x 15-5/8	55	1.125	200	80.00	36.0
1/2 Height	7-5/8 x 3-5/8 x 15-5/8		2.25		Lt Wt	27.0
1/2 Length	7-5/8 x 7-5/8 x 7-5/8		2.25			
6" Plain	5-5/8 x 7-5/8 x 15-5/8	45	1.125	210	60.00	28.0
1/2 Height	5-5/8 x 3-5/8 x 15-5/8		2.25		Lt Wt	21.0
1/2 Length	5-5/8 x 7-5/8 x 7-5/8		2.25			
4" Plain	3-5/8 x 7-5/8 x 15-5/8	35	1.125	210	50.00	24.0
1/2 Height	3-5/8 x 3-5/8 x 15-5/8		2.25		Lt Wt	18.0
1/2 Length	3-5/8 x 7-5/8 x 7-5/8		2.25			
16" Plain	15-5/8 x 7-5/8 x 15-5/8	70	2.25	170	150.00	61.0
14" Plain	13-5/8 x 7-5/8 x 15-5/8	65	2.25	180	90.00	56.0
10" Plain	9-5/8 x 7-5/8 x 15-5/8	60	2.25	195	85.00	43.0

Average 2 cubic yards mortar per M Block.

TABLE B

				Modular Manufactured Size - Inches			Mortar CuFt	Unit Per	Units Placed	Haul & Unload	Unit
				l	h	d	M	SqFt	ManDay	M PCs	Wt.
0403.0	**CLAY BACKING & PARTITION TILE**										
	3"			3 x	11-1/2 x	11-1/2	18.7	1.00	260	25.0	13.0
	4"			4 x	11-1/2 x	11-1/2	22.2	1.00	250	29.0	15.0
	6"			6 x	11-1/2 x	11-1/2	16.0	1.00	230	38.0	20.0
0404.0	**CLAY FACING TILE (GLAZED)**										
	2"	6T		1-3/4 x	5-1/16 x	7-3/4	13	2.25	180	20.00	6.5
	4"	6T		3-3/4 x	5-1/16 x	11-3/4		2.25		30.00	10.5
	6"	6T		5-3/4 x	5-1/16 x	11-3/4		2.25		45.00	15.2
	8"	6T		7-3/4 x	5-1/16 x	11-3/4		2.25		55.00	19.0
	2"	8W		1-3/4 x	7-3/4 x	15-3/4	22	1.13	125	40.00	14.0
	4"	8W		3-3/4 x	7-3/4 x	15-3/4		1.13		60.00	21.0
	6"	8W		5-3/4 x	7-3/4 x	15-3/4		1.13		85.00	31.0
0405.0	**GLASS UNITS**										
	6" x 6"			5-3/4 x	5-3/4 x	3-7/8	12.5	4.00	140	18.00	4.0
	6" x 8"			7-3/4 x	7-3/4 x	3-7/8	16.0	2.25	125	26.00	7.0
	12" x 12"			11-3/4 x	11-3/4 x	3-7/8	23.3	1.00	90	33.00	16.0
0406.0	**TERRA-COTTA**										
	3"			3" x	12" x	12"	25	1.00		36.00	10.0
	4"			4" x	12" x	12"	30	1.00		40.00	13.0
	6"			6" x	12" x	12"	40	1.00		47.00	18.0
	8"			8" x	12" x	12"	60				
0407.0	**MISCELLANEOUS UNITS**										
	Coping Tile			9"			50	-	160	35.00	-
				12"			60	-	140	40.00	-
	Flue Lining			8" x	8"		50	-	160	35.00	-
				12" x	12"		55	-	130	45.00	-
				18" x	18"		65	-	60	55.00	-
				21" x	21"		75	-	40	70.00	-
				24" x	24"		90	-	30	90.00	-
0408.0	**NATURAL STONE**										
	Cut Stone						SqFt				CuFt
	Lime Stone						10.0	-	18	25.00	150.0
	Marble						10.0	-	13	25.00	160.0
	Granite						10.0	-	11	25.00	165.0
	Ashlar Stone						15.0	-	26	25.00	150.0
	Rough Stone - Flagstone						15.0	-	22	25.00	140.0
	Field Stone						15.0	-	22	25.00	160.0
	Light Weight Stone						15.0	-	24	25.00	220.0
0409.0	**PRECAST VENEERS**										
	Architectural Precast Stone						10.0	-	18	25.00	150.0
	Terra-Cotta						10.0	-	17	25.00	130.0
	Precast Concrete						10.0	-	24	25.00	150.0

DIVISION #5 - METALS

Wage Rates (Including Fringes) & Location Modifiers
July 2001-2002

	Metropolitan Area		Iron Workers Wage Rate	Wage Rate Location Modifier
1.	Akron	*	42.04	129
2.	Albany-Schenectady-Troy		32.84	102
3.	Atlanta		25.71	80
4.	Austin	**	22.74	71
5.	Baltimore		34.16	95
6.	Birmingham	**	23.60	75
7.	Boston		40.19	127
8.	Buffalo-Niagara Falls		36.46	113
9.	Charlotte	**	20.75	64
10.	Chicago-Gary		45.03	137
11.	Cincinnati		30.74	98
12.	Cleveland		37.83	115
13.	Columbus	*	32.79	102
14.	Dallas-Fort Worth	**	21.00	67
15.	Dayton	*	31.98	100
16.	Denver-Boulder		24.33	80
17.	Detroit		42.00	128
18.	Flint	*	33.32	105
19.	Grand Rapids		33.52	105
20.	Greensboro-West Salem	**	20.75	64
21.	Hartford-New Britain	*	33.89	107
22.	Houston	**	21.95	76
23.	Indianapolis		33.48	106
24.	Jacksonville	**	23.29	73
25.	Kansas City		32.09	101
26.	Los Angeles-Long Beach		38.42	124
27.	Louisville	**	27.66	88
28.	Memphis	**	24.59	80
29.	Miami	**	24.22	76
30.	Milwaukee		36.30	113
31.	Minneapolis-St. Paul		37.27	111
32.	Nashville	**	23.24	75
33.	New Orleans	**	22.57	69
34.	New York		63.37	202
35.	Norfolk-Portsmouth	**	23.86	77
36.	Oklahoma City	**	25.37	82
37.	Omaha-Council Bluffs		28.56	81
38.	Orlando	**	27.17	86
39.	Philadelphia		44.26	143
40.	Phoenix	**	28.63	100
41.	Pittsburgh		37.55	118
42.	Portland		33.87	112
43.	Providence-Pawtucket	*	36.28	115
44.	Richmond	**	23.34	77
45.	Rochester		33.14	109
46.	Sacramento		39.45	127
47.	St. Louis		37.48	107
48.	Salt Lake City-Ogden	**	24.68	77
49.	San Antonio	**	20.38	72
50.	San Diego		40.85	130
51.	San Francisco-Oakland-San Jose		40.64	128
52.	Seattle-Everett		62.64	106
53.	Springfield-Holyoke-Chicopee	*	34.26	114
54.	Syracuse	*	33.29	108
55.	Tampa-St. Petersburg	**	38.24	87
56.	Toledo		33.14	110
57.	Tulsa	**	26.15	82
58.	Tucson	**	31.15	100
59.	Washington D.C.		29.65	90
60.	Youngstown-Warren		32.33	104
	AVERAGE		**30.58**	

Impact Ratio : Labor 30%, Material 70%
* Contract Not Settled - Wage Interpolated
** Non Signatory or Open Shop Rate

		PAGE
0501.0	**STRUCTURAL METAL FRAMING (05100)**	**5-3**
.1	BEAMS and COLUMNS	5-3
.2	CHANNELS	5-3
.3	ANGLES	5-3
.4	STRUCTURAL T's	5-3
.5	STRUCTURAL TUBING	5-3
.6	PLATES and BARS	5-3
.7	BOX GIRDERS	5-3
.8	TRUSSES	5-3
0502.0	**MISCELLANEOUS METALS (05500)**	**5-3**
.1	BOLTS	5-3
.2	CASTINGS (05540)	5-3
.3	CORNER GUARDS	5-3
.4	FIRE ESCAPES	5-3
.5	FRAMES, CHANNEL	5-3
.6	GRATINGS and TRENCH COVERS (05532)	5-3
.7	GRILLES	5-3
.8	HANDRAILS and RAILINGS (05720)	5-3
.9	LADDERS (05515)	5-3
.10	LINTELS, LOOSE	5-3
.11	PIPE GUARDS	5-4
.12	PLATE, CHECKERED	5-4
.13	STAIRS - METAL PAN (05712)	5-4
.14	STAIRS - CIRCULAR (05715)	5-4
.15	WIRE GUARDS	5-4
0503.0	**ORNAMENTAL METALS - NON FERROUS (05760)**	**5-4**
0504.0	**OPEN WEB JOISTS (05160)**	**5-4**
.1	STANDARD BAR JOISTS	5-4
.2	LONG SPAN JOISTS	5-4
0505.0	**METAL DECKINGS (05300)**	**5-4**
.1	SHORT SPAN	5-4
.2	LONG SPAN	5-4
.3	CELLULAR - ELECTRICAL and AIR FLOW	5-5
.4	CORRUGATED	5-5
.5	BOX RIB	5-5
0506.0	**METAL SIDINGS AND ROOFINGS (07411)**	**5-5**
.1	INDUSTRIAL	5-5
.11	Steel	5-5
.12	Aluminum	5-5
.2	ARCHITECTURAL	5-5
.21	Steel	5-5
.22	Aluminum	5-5
.23	Stainless Steel	5-5
.24	Protected Steel	5-5
	See Section 1310 for Metal Buildings	
0507.0	**LIGHT GAGE COLD-FORMED METAL FRAMING (05400)**	**5-5**
0508.0	**EXPANSION CONTROL (05800)**	**5-5**
	See Section 0303 for Reinforcing Steel	

0501.0 STRUCTURAL METAL (L+M) (Ironworkers)

		UNIT	LABOR	MATERIAL
.1	BEAMS AND COLUMNS			
.11	Wide Flange Shapes, H-Bearing Piles		(Subcontracted)	
	Light Weight Flange Columns & Misc. Columns			
	Light Beams - American Standard			
	Small Job (5 to 30 ton)	Ton	407.00	1,300.00
	Medium Job (30 to 100 ton)	Ton	397.00	1,280.00
	Large Job (100 ton and up)	Ton	376.00	1,230.00
.12	Junior Beams	Ton	386.00	1,330.00
.2	CHANNELS - American Std. - Junior -3" to 15"	Ton	525.00	1,200.00
	10" to 12"	Ton	505.00	1,330.00
.3	ANGLES (Equal and Unequal)	Ton	577.00	1,450.00
.4	STRUCTURAL T's (from wide Flange Shapes)			
	Light Shapes and American Standard	Ton	407.00	1,280.00
.5	STRUCTURAL TUBING (Steel Pipe, Square & Rectangular)	Ton	417.00	1,300.00
.6	PLATES AND BARS	Ton	438.00	1,365.00
.7	BOX GIRDERS	Ton	721.00	1,300.00
.8	TRUSSES	Ton	742.00	1,760.00
	Add for Welded Field Connections	Ton	165.00	-
	Add for High Strength Bolting	Ton	-	175.00
	Add for Riveted	Ton	319.00	-
	Add for Weathering Steel	Ton	-	250.00
	Add for Galvanizing	Ton	-	340.00

0502.0 MISCELLANEOUS METALS (L+M) (Ironworkers)

		UNIT	LABOR	MATERIAL
.1	BOLTS (with Nuts & Washers in Concrete or Masonry) See 0201.36 for Hole Drilling in Field			
	1/2" x 4"	Each	2.00	1.50
	5/8" x 4"	Each	2.10	1.75
	3/4" x 4"	Each	2.30	2.25
	1" x 4"	Each	2.45	3.00
.2	CASTINGS			
	Manhole Ring & Cover -24" Light Duty	Each	72.10	240.00
	24" Heavy Duty	Each	82.40	295.00
	Wheel Guards- 2' 6"	Each	77.25	180.00
.3	CORNER GUARDS- 3" x 3" x 1/4" Steel	LnFt	4.74	5.80
	4" x 4" x 1/4" Steel	LnFt	5.36	7.30
	Galv.	LnFt	5.36	8.80
.4	FIRE ESCAPES- 24" Steel Grate	Per Floor	906.00	2,500.00
		Tread	58.71	155.00
.5	FRAMES, CHANNEL- 6" - 8.2 lbs	LnFt	4.33	9.60
	8" - 11.5 lbs	LnFt	5.67	13.00
	10" - 15.3 lbs	LnFt	7.62	23.00
.6	GRATINGS- Steel - 1" x 1/8"	SqFt	1.49	6.90
	Aluminum - 1" x 1.8"	SqFt	1.49	14.00
	Cast Iron #2 - Light Duty	SqFt	1.49	9.60
	Heavy Duty	SqFt	1.75	18.00
.7	GRILLES	SqFt	3.09	8.10
.8	HANDRAILS - 1 1/2"			
	Wall Mounted - Single Rail- Steel	LnFt	5.77	9.50
	Aluminum	LnFt	6.59	35.50
	Floor Mounted - 2 Rail- Steel	LnFt	7.67	21.00
	3 Rail - Steel	LnFt	8.45	26.00
	Add for Galvanizing	LnFt	-	20%
.9	LADDERS - Steel	LnFt	10.30	33.00
	Aluminum	LnFt	9.99	47.00
.10	LINTELS, LOOSE	Ton	474.00	1030.00
	3" x 3" x 1/4" - 4.9 lbs per foot	LnFt	1.65	3.60
	4" x 3" x 1/4" - 5.8 lbs per foot	LnFt	1.96	4.10
	4" x 3 1/2" x 1/4" - 6.2 lbs per foot	LnFt	2.11	4.40
	4" x X3 1/2" x 5/16" - 7.7 lbs per foot	LnFt	2.68	5.20
	4" x X3 1/2" x 3/8" - 9.1 lbs per foot	LnFt	3.09	6.10
	4" x 4" x 1/4" - 6.6 lbs per foot	LnFt	2.47	4.75
	Add for Punching Holes	Each	-	3.30
	Add for Galvanizing	LnFt	-	1.00

		UNIT	LABOR	MATERIAL (Subcontracted)
0502.0	**MISCELLANEOUS METALS (Cont'd...)**			
.11	PIPE GUARDS - 4" - 4' Above Grade &	LnFt	8.86	6.43
	(Bollards in Concrete) - 4' Below Grade	LnFt	9.37	8.73
.12	PLATE, CHECKERED - 1/4" - 10.2 lbs	SqFt	1.55	5.92
	1/8" - 6.2 lbs	SqFt	1.29	5.00
.13	STAIRS, METAL PAN - 4' Wide x 10' (No Concrete)	PerFl	793.00	3,009.00
	Including Supports	Tread	56.65	194.00
.14	STAIRS, CIRCULAR - 3' x 10' - Aluminum	PerFl	448.00	3,060.00
	Steel	PerFl	494.00	3,264.00
	Cast Iron	PerFl	536.00	3,672.00
.15	WIRE GUARDS	SqFt	2.16	7.96
.16	TRENCH FRAME AND COVER - 2'	LnFt	5.56	9.28
0503.0	**ORNAMENTAL METALS (M) (Ironworkers)**			
	Non Ferrous - See 0502.0			
0503.0	**OPEN WEB JOISTS (L+M) (Ironworkers)**			
.1	STANDARD BAR JOISTS (Spans to 48')			
	Small Jobs (to 20 ton) - Spans to 15'	Ton	319.00	928.00
	25'	Ton	324.00	887.00
	35'	Ton	309.00	857.00
	Medium Jobs (20 - 50 ton) - Spans to 15'	Ton	309.00	908.00
	25'	Ton	304.00	872.00
	35'	Ton	299.00	734.00
	48'	Ton	288.00	821.00
	Large Jobs (50 ton & up) - Spans to 15'	Ton	299.00	893.00
	25'	Ton	288.00	867.00
	35'	Ton	288.00	847.00
	48'	Ton	278.00	836.00
.2	LONG SPAN JOISTS (Spans 48' to 150')			
	Small Jobs (to 30 ton) - Spans to 60'	Ton	319.00	903.00
	Medium Jobs (30 -100 ton) - Spans to 60'	Ton	314.00	893.00
	80'	Ton	299.00	918.00
	100'	Ton	278.00	928.00
	Large Jobs (100 ton & Up) - Spans to 60'	Ton	299.00	877.00
	80'	Ton	294.00	908.00
	100'	Ton	288.00	938.00
	120'	Ton	278.00	1,020.00
	150'	Ton	273.00	1,071.00
0505.0	**METAL DECKING (L+M) (Sheet Metal Workers)**			
.1	SHORT SPAN (Spans to 6')			
	(Baked Enamel - Ribbed, V Beam & Seam)			
	1 1/2" Deep - 18 Ga	Sq	47.38	130.00
	20 Ga	Sq	46.35	102.00
	22 Ga	Sq	44.29	91.80
	Add for Galvanized	Sq	-	10%
.2	LONG SPAN (Spans to 12') (Enameled & Open Bottom)			
	3" Deep - 16 Ga	Sq	63.86	388.00
	18 Ga	Sq	59.74	342.00
	20 Ga	Sq	57.68	301.00
	22 Ga	Sq	55.62	245.00
	4 1/2" Deep - 16 Ga	Sq	69.01	479.00
	18 Ga	Sq	67.98	423.00
	20 Ga	Sq	65.92	383.00
	Add for Acoustical	Sq	-	20%
	Add for Galvanized	Sq	-	20%

		UNIT	LABOR (Subcontracted)	MATERIAL
0505.0	**METAL DECKING (L+M) (Sheet Metal), Cont'd...**			
.3	CELLULAR - Electric and Air Flow			
	3" Deep - 16 - 16 Ga Galvanized	Sq	75.00	574.00
	18 - 18 Ga Galvanized	Sq	70.00	533.00
	4 1/2" Deep- 16 - 16 Ga Galvanized	Sq	88.00	702.00
	18 - 18 Ga Galvanized	Sq	82.00	584.00
.4	CORRUGATED METAL			
	Standard	*Spans to Approx.*		
	Black - .015" — 3' 6"	Sq	36.00	59.00
	Galvanized - .015" — 3' 6"	Sq	36.00	72.00
	Heavy Duty			
	Black - 26 Ga — 5' 0"	Sq	39.00	69.00
	Galvanized - 26 Ga — 5' 0"	Sq	39.00	85.00
	Super Duty			
	Black - 24 Ga — 7' 0"	Sq	41.00	98.00
	Galvanized - 24 Ga — 7' 0"	Sq	41.00	111.00
	Black - 22 Ga — 7' 6"	Sq	42.00	119.00
	Galvanized - 22 Ga — 7' 6"	Sq	42.00	123.00
.5	BOX RIB			
	2" - 24 Ga	Sq	56.00	364.00
	22 Ga	Sq	60.00	384.00
	20 Ga	Sq	64.00	410.00
0506.0	**METAL SIDINGS & ROOFINGS (L+M) (Sheet Metal)**			
.1	INDUSTRIAL			
.11	Steel (24 Ga - See 0505 for Other Gauges)			
	Corrugated - Galvanized	Sq	70.00	103.00
	Ribbed - Enameled	Sq	64.00	98.00
	Add for Liner Panel	Sq	32.00	92.00
	Add for Insulation	Sq	27.00	29.00
.12	Aluminum- Corrugated	Sq	60.00	123.00
.2	ARCHITECTURAL (Prefinished)			
.21	Steel - Baked Enameled	Sq	93.00	164.00
	Porcelain Enameled	Sq	118.00	205.00
	Acrylic Enameled	Sq	113.00	181.00
	Plastic Faced	Sq	113.00	246.00
.22	Aluminum - Anodized	Sq	110.00	185.00
	Plastic Faced	Sq	112.00	246.00
	Porcelainized	Sq	115.00	236.00
	Add for Liner Panel and Insulation	Sq	43.00	108.00
.23	Stainless Steel	Sq	180.00	636.00
.24	Protected Metal (Asphalt on Steel w/Finish)	Sq	134.00	354.00
	See 1310 for Metal Buildings			
0507.0	**LIGHT GAGE FRAMING (L+M) (Ironworkers)**			
	3 5/8"- 16 Ga	LnFt	.58	2.92
	6"- 16 Ga	LnFt	.61	3.69
	8"- 16 Ga	LnFt	.74	4.31
	10"- 16 Ga	LnFt	.89	5.43
	Average - Wall	Lb	.34	.77
	Average - Hanging	Lb	.70	.92
0508.0	**EXPANSION CONTROL (L+M) (Ironworkers)**			
	4"- Aluminum	LnFt	6.70	37.00
	Bronze	LnFt	7.21	95.00
	Stainless Steel	LnFt	7.21	98.00
	2"- Aluminum	LnFt	5.41	28.00
	Bronze	LnFt	6.70	56.00
	Stainless Steel	LnFt	6.70	62.00

		Sq. Ft. Cost
0501.0	**Structural Steel Frame**	
	To 30 Ton 20' Span	6.15
	24' Span	6.61
	28' Span	7.18
	32' Span	7.89
	36' Span	8.71
	40' Span	9.43
	44' Span	10.05
	48' Span	10.97
	52' Span	12.30
	56' Span	13.33
	60' Span	14.15
	Deduct for Over 30 Ton	5%
0504.0	**Open Web Joists**	
	To 20 Ton 20' Span	1.79
	24' Span	1.85
	28' Span	1.90
	32' Span	2.00
	36' Span	2.15
	40' Span	2.46
	48' Span	2.92
	52' Span	4.05
	56' Span	3.49
	60' Span	4.10
	Deduct for Over 20 Ton	10%
0505.0	**Metal Decking**	
	1 1/2" Deep - Ribbed - Baked Enamel - 18 Ga	2.05
	20 Ga	1.95
	22 Ga	1.74
	3" Deep - Ribbed - Baked Enamel - 18 Ga	4.51
	20 Ga	4.10
	22 Ga	3.54
	4 1/2" Deep - Ribbed - Baked Enamel - 16 Ga	6.15
	18 Ga	5.23
	20 Ga	4.61
	3" Deep - Cellular - 18 Ga	7.48
	16 Ga	8.30
	4 1/2" Deep - Cellular - 18 Ga	8.51
	16 Ga	10.05
	Add for Galvanized	10%
	Corrugated Black Standard .015	1.08
	Heavy Duty - 26 Ga	1.18
	S. Duty - 24 Ga	1.64
	22 Ga	1.74
	Add for Galvanized	15%
0506.0	**Metal Sidings and Roofings**	
	Aluminum- Anodized	3.38
	Porcelainized	7.07
	Corrugated	1.85
	Enamel, Baked Ribbed	2.97
	24 Ga Ribbed	2.00
	Acrylic	3.49
	Porcelain Ribbed	3.95
	Galvanized- Corrugated	1.90
	Plastic Faced	4.25
	Protected Metal	5.13
	Add for Liner Panels	1.54
	Add for Insulation	.67

DIVISION #6 - WOOD & PLASTICS

Wage Rates (Including Fringes) & Location Modifiers
July 2001-2002

	Metropolitan Area		Carpenter Rate	Wage Rate Location Modifier
1.	Akron	*	32.24	114
2.	Albany-Schenectady-Troy		28.43	102
3.	Atlanta		21.16	80
4.	Austin	**	21.14	74
5.	Baltimore		25.09	92
6.	Birmingham	**	21.13	75
7.	Boston		37.87	136
8.	Buffalo-Niagara Falls		36.90	128
9.	Charlotte	**	19.26	67
10.	Chicago-Gary	*	37.56	126
11.	Cincinnati		26.80	95
12.	Cleveland		32.54	117
13.	Columbus	*	27.60	99
14.	Dallas-Fort Worth	**	19.19	68
15.	Dayton	*	30.08	105
16.	Denver-Boulder		23.14	81
17.	Detroit		36.71	127
18.	Flint	*	29.18	100
19.	Grand Rapids		29.17	100
20.	Greensboro-West Salem	**	19.36	67
21.	Hartford-New Britain	*	32.91	116
22.	Houston	**	22.21	73
23.	Indianapolis		30.03	104
24.	Jacksonville	**	23.64	82
25.	Kansas City		28.58	98
26.	Los Angeles-Long Beach		31.72	122
27.	Louisville	**	23.43	82
28.	Memphis	**	19.93	87
29.	Miami	**	21.37	75
30.	Milwaukee	*	33.10	112
31.	Minneapolis-St. Paul		33.24	113
32.	Nashville	**	22.72	79
33.	New Orleans	**	20.76	73
34.	New York		53.52	193
35.	Norfolk-Portsmouth	**	22.34	79
36.	Oklahoma City	**	21.37	75
37.	Omaha-Council Bluffs		22.25	75
38.	Orlando	**	30.08	89
39.	Philadelphia	*	39.20	135
40.	Phoenix	**	23.83	95
41.	Pittsburgh	*	31.22	109
42.	Portland	*	32.23	113
43.	Providence-Pawtucket	*	32.34	114
44.	Richmond	**	22.30	78
45.	Rochester		33.61	116
46.	Sacramento	*	39.04	132
47.	St. Louis		34.20	117
48.	Salt Lake City-Ogden	**	22.11	70
49.	San Antonio	**	20.74	75
50.	San Diego	*	32.38	124
51.	San Francisco-Oakland-San Jose		39.04	130
52.	Seattle-Everett		33.55	116
53.	Springfield-Holyoke-Chicopee	*	31.38	110
54.	Syracuse	*	27.47	114
55.	Tampa-St. Petersburg	**	25.60	91
56.	Toledo	*	33.09	117
57.	Tulsa	**	21.78	75
58.	Tucson	**	27.01	95
59.	Washington D.C.	*	23.77	87
60.	Youngstown-Warren	*	29.22	107
	AVERAGE		**28.38**	

Impact Ratio: Labor 50%, Material 50%
*   Contract Not Settled - Wage Interpolated
** Non Signatory or Open Shop Rate

			PAGE
0601.0	**ROUGH CARPENTRY (CSI 06100)**		**6-3**
.1	LIGHT FRAMING AND SHEATHING		6-3
.11		Joists, Plates and Framing	6-3
.12		Studs, Plates and Framing	6-3
.13		Bridging	6-3
.14		Rafters	6-3
.15		Stairs	6-3
.16		Subfloor Sheathing	6-4
.17		Floor Sheathing	6-4
.18		Wall Sheathing	6-4
.19		Roof Sheathing	6-4
.2	HEAVY FRAMING AND DECKING		6-5
.21		Columns and Posts	6-5
.22		Decking	6-5
.3	MISCELLANEOUS CARPENTRY		6-6
.31		Blocking and Bucks	6-6
.32		Grounds, furring, Sleepers and nailers	6-6
.33		Cant Strips	6-6
.34		Building Papers and Sealers	6-6
0602.0	**FINISH CARPENTRY (CSI 006200)**		**6-7**
.1	FINISH SIDINGS AND FACING MATERIALS (Exterior)		6-7
.11		Boards, Beveled, and Lap Siding	6-7
.12		Plywood - Siding	6-7
.13		Hardboard - Paneling and lap Siding	6-7
.14		Shingles	6-7
.15		Facias and Trim	6-8
.16		Soffits	6-8
.2	FINISH WALLS AND CEILING MATERIALS (Interior)		6-8
.21		Boards	6-8
.22		Hardboard or Pressed Wood (4' x 8' Panels)	6-8
.23		Plywood Paneling (4' x 8' Panels and Prefinished)	6-8
.24		Gypsum Board and Metal Framing	6-9
0603.0	**MILLWORK AND CUSTOM WOODWORK**		**6-10**
.1	CUSTOM CABINET WORK		6-10
.2	COUNTER TOPS		6-10
.3	CUSTOM DOORS AND FRAMES		6-11
.4	MOULDINGS AND TRIM		6-11
.5	STAIRS		6-12
.6	SHELVING		6-12
.7	CUSTOM PANELING		6-12
0604.0	**GLUE LAMINATED (CSI 06180)**		**6-12**
.1	ARCHES AND RIGID FRAMES		6-12
.2	BEAMS, JOISTS AND PURLINS		6-12
.3	DECKING		6-12
0605.0	**PREFABRICATED WOOD AND PLYWOOD COMPONENTS (CSI 06170)**		**6-12**
.1	WOOD TRUSSED RAFTERS		6-12
.2	WOOD FLOOR AND ROOF TRUSS JOISTS		6-12
.21		Open Wood Web - Gusset Plate Connection	6-12
.22		Open Metal Web - Pin Connection	6-12
.23		Plywood Web	6-13
.3	LAMINATED VENEER STRUCTURAL BEAMS		6-13
.31		Micro Lam - Plywood	6-13
.32		Glue Lam - Dimension	6-13
0606.0	**WOOD TREATMENT (CSI 06300)**		**6-13**
0607.0	**ROUGH HARDWARE (CSI 06050)**		**6-13**
.2	NAILS, SCREWS, ETC.		6-13
.2	JOIST HANGERS		6-13
.3	BOLTS		6-13
0608.0	**EQUIPMENT**		**6-13**
6A	QUICK ESTIMATING		6A-14 thru 6A-19

0601.0 ROUGH CARPENTRY (Carpenters)

		UNIT	LABOR	MATERIAL
.1	LIGHT FRAMING & SHEATHING (Standard - Hem-Fir)			
.11	Joists, Beams and Headers - 16" O.C.			
	2"x 6" Floor	BdFt	.53	.53
	2"x 8" Floor	BdFt	.52	.54
	2"x10" Floor	BdFt	.50	.59
	2"x12" Floor	BdFt	.52	.62
	Add for Ceiling Joists - 2nd Fl or Roof	BdFt	.09	-
	Add for Door, Window & Misc. Headers	BdFt	.36	-
	Add for Sloped Installations	BdFt	40%	-
	Add for Ledgers	BdFt	.24	-
.12	Studs, Plates and Miscellaneous Framing - 16" O.C.			
	2" x 2"	BdFt	.77	.58
	2" x 3"	BdFt	.72	.56
	2" x 4"	BdFt	.62	.52
	2" x 6"	BdFt	.60	.53
	Add for 2nd Floor and Above per Floor	BdFt	.10	-
	Add for Bolted Plates or Sills	BdFt	.52	-
	Add for Each 1' Above 8' in Length	BdFt	.04	-
	Add for Fire Stops, Fillers, and Nailers	BdFt	.33	-
	Add for Soffit & Suspended Framing 2"x 4"	BdFt	.77	-
	See 0602 for Metal Studs and Plates			
.13	Bridging - 16" O.C.			
	1" x 3" Wood Diagonal	Set	1.39	.40
	2" x 4" Diagonal	Set	1.60	1.00
	1" Metal Diagonal - 18 Ga	Set	1.39	.95
	2" x 8" Solid	BdFt	1.65	.65
.14	Rafters and Field-Made Trussed Rafters - 16" O.C.			
	(See 0605 for Prefabrication)			
	2" x 4"	BdFt	.81	.52
	2" x 6"	BdFt	.79	.53
	2" x 8"	BdFt	.76	.54
	2" x 10"	BdFt	.71	.62
	2" x 4" Trusses - Make up in Field	BdFt	.62	.72
	Erection	BdFt	.43	-
	Add for Hips and Valleys	BdFt	.43	-
	Add for Bracing	BdFt	.82	.60
.15	Stairs			
	2" x 10" and 2" x 12" for Stringers	BdFt	1.75	.63
	1" x 8" and 2" x 10" for Risers & Stairs	BdFt	1.24	.64
	Average for Stringers, Treads and Risers	BdFt	1.49	.75
	Add to Above Framing Where Applicable:			
	Circular Installations	BdFt	60%	-
	Diagonal Installations	BdFt	10%	7%
	Kiln Dry Lumber	BdFt	-	15%
	Construction Grade Lumber	BdFt	-	.06
	Lumber Over 20'	BdFt	-	.25
	Winter Construction	BdFt	10%	-
	Add for Pressure Treating			
	2" x 4" - 2" x 6" - 2" x 8"	BdFt	-	.15
	2" x 10" - 2" x 12"	BdFt	-	.24
	Add for Fire Treating	BdFt	-	.25

0601.0   ROUGH CARPENTRY, Cont'd...

	UNIT	LABOR	MATERIAL
.16 Sub Floor Sheathing (Structural)			
1" x 8" and 1" x 10" #3 Pine	BdFt	.37	.57
1/2" x 4' x 8' Plywood - CD Exterior	SqFt	.35	.42
5/8" x 4' x 8' Plywood - CD Exterior	SqFt	.36	.50
3/4" x 4' x 8' Plywood - CD Exterior	SqFt	.37	.62
3/4" x 4' x 8' Plywood - T&G, CD Exterior	SqFt	.43	.85
5/8" x 4' x 8' Particle Board	SqFt	.37	.42
3/4" x 4' x 8' Particle Board	SqFt	.38	.50
3/4" x 4' x 8' Particle Board, T&G	SqFt	.42	.62
.17 Floor Sheathing (Over Sub Floor)			
With Partitions in Place			
1/4" x 4' x 8' Plywood - CD	SqFt	.38	.35
3/8" x 4' x 8' Plywood - CD	SqFt	.39	.36
1/2" x 4' x 8' Plywood - CD	SqFt	.40	.47
5/8" x 4' x 8' Plywood - CD	SqFt	.41	.59
3/8" x 4' x 8' Particle Board	SqFt	.39	.25
1/2" x 4' x 8' Particle Board	SqFt	.40	.30
5/8" x 4' x 8' Particle Board	SqFt	.41	.42
3/4" x 4' x 8' Particle Board	SqFt	.42	.52
.18 Wall Sheathing			
1" x 8" and 1" x 10" #3 Pine	BdFt	.42	.58
3/8" x 4' x 8' Plywood - CD Exterior	SqFt	.41	.33
1/2" x 4' x 8' Plywood - CD Exterior	SqFt	.42	.40
5/8" x 4' x 8' Plywood - CD Exterior	SqFt	.43	.48
3/4" x 4' x 8' Plywood - CD Exterior	SqFt	.45	.60
1/2" x 4' x 8' Fiberboard Impregnated	SqFt	.40	.25
25/32" x 4' x 8' Fiberboard Impregnated	SqFt	.41	.28
1" x 2' x 8' T&G Styrofoam	SqFt	.43	.32
2" x 2' x 8' T&G Styrofoam	SqFt	.57	.62
1/2" x 4' x 8' Wafer Board	SqFt	.41	.27
5/8" x 4' x 8' Wafer Board	SqFt	.44	.36
.19 Roof Sheathing - Flat			
1" x 6" and 1" x 8" #3 Pine	BdFt	.42	.57
1/2" x 4' x 8' Plywood - CD Exterior	SqFt	.39	.42
5/8" x 4' x 8' Plywood - CD Exterior	BdFt	.41	.50
3/4" x 4' x 8' Plywood - CD Exterior	SqFt	.42	.60
Add for Sloped Installation	SqFt	.12	-
Add Steep Sloped Installation (over 5-12)	SqFt	.28	-
Add for Dormer Work	SqFt	.52	-
Add to Above Sheathing where Applicable:			
Fire Treating	SqFt	-	.26
Pressure Treating	SqFt	-	.21
Winter Construction	SqFt	10%	-
Plywood - AC Exterior (Good One Side)	SqFt	-	.16
Plywood - AD Exterior (Good One Side)	SqFt	-	.18
Diagonal Installations	SqFt	8%	7%
Plywood 10' Lengths	SqFt	-	.18

Recommended 10% Decrease on Labor Units for Residential Construction

0601.0 ROUGH CARPENTRY, Cont'd...

.2 HEAVY FRAMING AND DECKING (Construction-Grade Fir, Air Dried S4S)
See 0604.0 for Laminated Heavy Framing & Decking.
See 0606.0 for Wood Treatments.
See 0109.8 for Wood Curbs and Walls.

	UNIT	LABOR	MATERIAL
.21 Columns and Posts			
4"x 4"	MBF	520.00	940.00
4"x 6"	MBF	510.00	960.00
6"x 6"	MBF	479.00	1,060.00
8"x 8"	MBF	474.00	1,260.00
12"x12"	MBF	453.00	1,700.00
Add for Fitting at Base and Heat	MBF	155.00	-
Beams and Joists			
4" x 4"	MBF	520.00	940.00
4" x 6"	MBF	510.00	960.00
4" x 8"	MBF	505.00	1,120.00
4" x10"	MBF	484.00	1,140.00
6" x 6"	MBF	489.00	1,120.00
6" x 8"	MBF	474.00	1,160.00
6" x10"	MBF	469.00	1,220.00
8" x12"	MBF	458.00	1,420.00
Add to 0601.21 Above for:			
Rough Sawn	MBF	-	129.00
Mill Cutting to Length and Shape	MBF	-	129.00
Nails	MBF	-	60.00
Hauling & Unloading (if Quoted Cars)	MBF	-	82.00
Equipment & Operator Cost - Fork Lift	MBF	-	62.00
Equipment & Operator Cost - Crane	MBF	-	103.00
Add for Red Cedar	MBF	-	1,100.00
Add for Redwood	MBF	-	1,300.00
.22 DECKING			
Tongue & Groove			
Fir, Hemlock & Spruce, Construction Grade (SPF)			
4" x 4"	MBF	335.00	940.00
2" x 6" and 2" x 8"	MBF	304.00	710.00
3" x 6"	MBF	314.00	860.00
4" x 6"	MBF	324.00	960.00
Cedar - #3 and better			
3" x 4"	MBF	314.00	1,500.00
4" x 4" - 4" x 6" - 4" x 8" and 4" x 10"	MBF	309.00	1,650.00
White Pine			
3" x 6" and 3" x 8"	MBF	314.00	1,500.00
4" x 6" and 4" x 8"	MBF	324.00	1,700.00
Add for D Grade	MBF	-	1,050.00
2" x 6" Panelized Decking - 1 1/2" x 20"			
Premium Grade	SqFt	.66	1.90
Architectural Grade	SqFt	.66	1.80
Add for Select Grade	MBF	-	5%
Add for Hemlock	MBF	-	45.00
Add for Hip Roof	MBF	62.00	-
Add for Specified Lengths	MBF	-	31.00
Add for End Matched	MBF	-	41.00
Add for Hauling and Unloading	MBF	-	62.00
Add for Nails	MBF	-	45.00

See 0604 for Laminated Decking
See 0604.3 for Square Foot Deck Costs

0601.0 ROUGH CARPENTRY, Cont'd...

		UNIT	LABOR	MATERIAL
.3	MISCELLANEOUS CARPENTRY			
.31	Blocking and Bucks (2" x 4" or 2" x 6")			
	Doors and Window Bucks			
	Erected Before Masonry	BdFt	1.29	.56
		or, Each	25.75	8.40
	Bolted to Masonry or Steel	BdFt	1.80	.56
		or, Each	36.05	8.40
	Nail Driven to Steel Studs or Masonry	BdFt	1.18	.56
		or, Each	23.69	8.40
	Roof Blocking - Pressure Treated			
	Edges - Bolted to Concrete	BdFt	1.34	.79
	Nailed to Wood	BdFt	.67	.79
	Openings - Nailed	BdFt	1.49	.79
	Add for 2" x 8" or 2" x 10"	BdFt	-	.11
	Add to 0601.31 for Bolts - 4' O.C.	BdFt	-	.32
.32	Grounds, Furring, Sleepers & Nailers #3 Pine			
	1" x 4" Fastened to Wood	BdFt	1.34	.90
		or, LnFt	.44	.30
	1" x 3" Fastened to Wood	LnFt	.41	.22
	1" x 2" Fastened to Wood	LnFt	.39	.16
	1" x 3" Fastened to Concrete/ Masonry			
	Nailed	LnFt	.56	.30
	Gun Driven (Incl Fasteners)	LnFt	.48	.35
	Pre Clipped	LnFt	.70	.32
	Add for Ceiling Work	LnFt	.15	-
	2" x 4" Not Suspended Framing	BdFt	1.34	.53
	Suspended Framing	BdFt	1.65	.53
	2" x 2" Not Suspended Framing	BdFt	1.70	.58
	Suspended Framing	BdFt	2.42	.58
.33	Cant Strips			
	Pressure Treated			
	Cut from 4" x 4" - At Roof Edges	BdFt	.41	.44
		or, LnFt	.56	.59
	4" x 4" - At Roof Openings	BdFt	.60	.44
		or, LnFt	.79	.59
	6" x 6" - At Roof Edges	BdFt	1.49	2.05
		or, LnFt	.98	1.37
	6" x 6" - At Roof Openings	BdFt	2.06	2.05
		or, LnFt	1.34	1.37
	Add for Bolting (Bolts - 4' O.C.)	BdFt	.62	.53
	Not Treated			
	Cut from 4" x 4" - At Roof Edges	BdFt	.41	.36
		or, LnFt	.56	.48
	6" x 6" - At Roof Edges	BdFt	1.49	1.75
		or, LnFt	.98	1.15
.34	Building Papers and Sealers			
	15" Felt (432 Sq.Ft. to Roll)	SqFt	.06	.03
	20" Felt (432 Sq.Ft. to Roll)	SqFt	.07	.04
	30" Felt (216 Sq.Ft. to Roll)	SqFt	.08	.06
	Cotton and Glass Asphalt Saturated	SqFt	.07	.06
	Polyethylene 4 mil	SqFt	.06	.02
	6 mil	SqFt	.07	.03
	30# Red Rosin Paper	SqFt	.07	.13
	Sill Sealer 6" x 100'	SqFt	.23	.21

0602.0 FINISH CARPENTRY (Carpenters)

.1 FINISH SIDINGS AND FACING MATERIAL (Exterior)
(See Quick Estimating for Square Foot Costs)

	UNIT	LABOR	MATERIAL
.11 Boards, Beveled and Lap Siding			
Cedar - Beveled - Clear 1/2" x 4"	BdFt	.98	1.47
1/2" x 6"	BdFt	.91	1.58
Beveled - Rough Sawn 7/8" x 8"	BdFt	.85	1.37
7/8" x 10"	BdFt	.81	1.58
7/8" x 12"	BdFt	.79	1.63
Board- Rough Sawn 3/4" x 12"	BdFt	.77	1.58
Redwood Beveled- Clear Heart 1/2" x 4"	BdFt	.98	1.63
1/2" x 6"	BdFt	.91	1.79
1/2" x 8"	BdFt	.85	1.89
5/8" x 10"	BdFt	.79	2.57
3/4" x 6"	BdFt	.83	2.42
3/4" x 8"	BdFt	.79	2.52
Board - Clear Heart 3/4" x 6"	BdFt	.76	2.99
3/4" x 8"	BdFt	.73	4.10
Board, - T&G 1" x 4" & 6"	BdFt	.79	3.26
Rustic Beveled - 1" x 8" & 10"	BdFt	.76	3.47
5/4"x 6" & 8"	BdFt	.73	3.36
Add for Metal Corners	Each	.80	.32
Add for Mitering Corners	Each	2.58	-
.12 Plywood - Siding			
Fir - AC - Smooth One Side 1/4"	SqFt	.52	.46
3/8"	SqFt	.54	.50
Fir - Rough Faced 3/8"	SqFt	.56	.58
Fir - AC - Smooth One Side 1/2"	SqFt	.56	.60
5/8"	SqFt	.60	.72
3/4"	SqFt	.64	.90
Fir - Grooved and Rough Faced 5/8"	SqFt	.62	.80
Cedar - Rough Sawn 3/8"	SqFt	.55	.92
5/8"	SqFt	.62	1.20
3/4"	SqFt	.65	1.35
Cedar - Grooved & Rough Faced (4" & 8") 5/8"	SqFt	.62	1.31
Add for 4' x 10' and 4' x 9' Sheets	SqFt	-	.14
Add for Wood Batten Strips 1" x 2" Pine	LnFt	.54	.29
Add for Horizontal Joint Flashing (Spline)	LnFt	.43	.37
Add for Medium Density Fir	SqFt	.02	.16
.13 Hardboard - Paneling and Lap Siding (Primed)			
7/16" - Rough Textured 4' x 8' Paneling	SqFt	.47	.93
7/16" - Grooved 4' x 8' Paneling	SqFt	.52	.93
7/16" - Stucco Board 4' x 8' Paneling	SqFt	.52	.98
Lap Siding - 7/16" x 8"	BdFt	.70	.67
7/16"x 12"	BdFt	.77	.72
Add for Prefinishing	BdFt	-	.15
.14 Shingles - Wood			
16" Red Cedar - 12" to Weather - #1	SqFt	.73	1.68
#2	SqFt	.70	1.31
#3	SqFt	.64	.95
Fancy Butt Red Cedar 7" #1	SqFt	.76	3.68
Red Cedar Hand Split - #1 1/2" to 3/4"	SqFt	.89	1.37
3/4" to 1 1/4" - #1	SqFt	.92	1.63
3/4" to 1 1/4" - #2	SqFt	.89	1.73
3/4" to 1 1/4" - #3	SqFt	.87	1.63
Add for Fire Retardant	SqFt	-	.42
Add for 3/8" Backer Board	SqFt	.27	.21
Add for Metal Corners	Each	.74	.32
Add for Ridges, Hips and Corners	LnFt	1.24	.84
Add for 8" to Weather	SqFt	25%	-

See Division 0708 for Roof Shingles

0602.0 FINISH CARPENTRY, Cont'd...

.15 Facias and Trim	UNIT	LABOR	Pine #2	Cedar #3	Redwood Clear	Plywood 5/8" ACX
All Sizes (average)	BdFt	.72	1.04	1.58	3.57	.75
1" x 3"	LnFt	.45	.30	.39	.84	.19
1" x 4"	LnFt	.52	.40	.47	1.21	.25
1" x 6"	LnFt	.60	.60	.84	1.73	.38
1" x 8"	LnFt	.65	.80	1.16	2.10	.50
1" x 12"	LnFt	.71	1.04	1.58	3.57	.75
Add for Clear Grades			400%	100%	-	30%

.16 Soffits	UNIT	LABOR	MATERIAL
1/2" Plywood Fir ACV	SqFt	.88	.60
Aluminum	SqFt	.88	1.20
Plastic - Egg Crate	SqFt	.88	1.30

.2 FINISH WALLS AND CEILING MATERIALS (Interior)

.21 Boards (See Quick Est. - Size/Cutting Allowance Added)

	UNIT	LABOR	MATERIAL
Cedar - #3 Grade - 1" x 6"	SqFt	.74	1.80
Knotty 1" x 6"	SqFt	.72	1.94
1" x 8"	SqFt	.69	2.10
D Grade 1" x 4"	SqFt	.74	2.63
1" x 6"	SqFt	.72	2.73
1" x 8"	SqFt	.70	2.63
Aromatic 1" x 6"	SqFt	.74	2.73
Redwood - Constr.1" x 6"	SqFt	.78	2.00
1" x 8"	SqFt	.72	2.21
Add for Clear	SqFt	10%	100%
Fir - Beaded 5/8" x 4"	SqFt	.78	1.80
#2 Grade 1" x 6"	SqFt	.73	2.00
1" x 8"	SqFt	.70	1.95
1" x 10"	SqFt	.66	2.00

.22 Hardboard & Pressed Woods (4' x 8' Panels)

	UNIT	LABOR	MATERIAL
Tempered 1/8"	SqFt	.48	.33
1/4"	SqFt	.53	.41
Pegboard (Perforated)1/8"	SqFt	.49	.37
1/4"	SqFt	.54	.50
Plastic Faced 1/4"	SqFt	.57	.61
Flake Board 1/8"	SqFt	.44	.35
Add for Metal or Plastic Mouldings	LnFt	.46	.37
Add for Pre-finished	SqFt	-	.16

.23 Plywood Paneling (4'x 8' Panels, Prefinished)

	UNIT	LABOR	MATERIAL
Birch, Natural 1/4"	SqFt	.76	.95
3/4"	SqFt	.87	1.58
White Select 1/4"	SqFt	.77	1.47
Oak, Red - Rotary Cut 1/4"	SqFt	.77	.89
3/4"	SqFt	.88	1.89
White 1/4"	SqFt	.76	1.89
Cedar, Aromatic 1/4"	SqFt	.77	1.58
Cherry 3/4"	SqFt	.88	2.21
Pecan 3/4"	SqFt	.88	2.21
Walnut 1/4"	SqFt	.76	2.21
3/4"	SqFt	.88	3.36
Mahogany, Lauan 1/4"	SqFt	.76	.63
3/4"	SqFt	.88	1.05
African 3/4"	SqFt	.88	2.31
Knotty Pine 1/4"	SqFt	.76	1.05
3/4"	SqFt	.88	1.68
Slat Wall - Painted 3/4"	SqFt	.88	1.42
Oak 3/4"	SqFt	.93	2.21
Add for V Groove and T&G	SqFt	-	.26
Add for Glued-On Applications	SqFt	.15	.07
Add for 10' Lengths	SqFt	.10	.21

0602.0   FINISH CARPENTRY, Cont'd...

.24   Gypsum Board and Framing
(See 0902 - Subcontracted Work Costs)
(See 0902 - Metal Framing SqFt Costs)

	UNIT	LABOR	MATERIAL
Finish Board (Walls to 8' - Screwed On)			
3/8" 4' x 8' and 12'	SqFt	.30	.26
1/2" 4' x 8' and 12'	SqFt	.31	.27
5/8" 4' x 8' and 12	SqFt	.32	.29
1" 4' x 8' and 12' Core Board	SqFt	.33	.46
1" 4' x 8' and 12' Bead Board	SqFt	.31	.46
Add for Adhesive Application	SqFt	.06	.02
Deduct for Nail-on Application	SqFt	.02	-
Add for: Fire Resistant Board	SqFt	-	.01
Add for: Foil Backed (Insulating)	SqFt	-	.09
Add for: Moisture Resistant Type 5/8" & 1/2"	SqFt	-	.05
Add for: Work Above 8'	SqFt	.06	-
Add for: Work Over 2-Story per Story	SqFt	.05	-
Add for: Ceiling Work - To Wood	SqFt	.11	-
To Steel	SqFt	.30	-
Add for: Resilient Clip Application	SqFt	.29	.06
Add for: Small Cut Up Areas	SqFt	.32	.03
Add for: Beam and Column Work	SqFt	.41	.03
Add for: Circular Construction	SqFt	.35	.03
Add for: Vinyl Faced Board			
3/8" x 4' x 8' - Cost Variable	SqFt	.35	.65
1/2" x 4' x 8' - Cost Variable	SqFt	.39	.70
Metal Framing (Studs, Runners and Channels)			
Studs 25-Gauge 1 5/8"	LnFt	.45	.20
2 1/2"	LnFt	.46	.21
3 5/8"	LnFt	.52	.23
6"	LnFt	.48	.32
8"	LnFt	.52	.35
Runners or Track 1 5/8"	LnFt	.43	.20
2 1/2"	LnFt	.45	.21
3 5/8"	LnFt	.47	.23
6"	LnFt	.48	.30
8"	LnFt	.52	.31
7/8" Furring Channels	LnFt	.45	.22
1 1/2" Furring Channels (Cold Rolled)	LnFt	.46	.28
Resilient Channels	LnFt	.43	.21
Add for Over 10' Lengths	LnFt	.16	.07
Add for 20-Gauge 6" Material	LnFt	.15	.40
Add for 16-Gauge 8" Material	LnFt	.26	.90
Metal Trim			
Casing Bead	LnFt	.62	.15
Corner Bead	LnFt	.47	.09
J Bead 3/8" - 1/2" - 5/8"	LnFt	.47	.15
L Bead 1/2" - 5/8"	LnFt	.47	.15
Expansion Bead - Metal	LnFt	.47	.50
Plastic	LnFt	.47	.25
Metal Mouldings	LnFt	.72	.35
Taping and Sanding	SqFt	.26	.05
Texturing	SqFt	.28	.07
Thin Coat Plaster - 1 Coat	SqFt	.77	.25
Sound Deadening Insulation - 2 1/2"	SqFt	.27	.22
3 1/2"	SqFt	.31	.24
6"	SqFt	.33	.31

Recommend 10% decrease on Labor Units for Residential Construction

0603.0 MILLWORK & CUSTOM WOODWORK (M) (Carpenters)

	UNIT	LABOR	MATERIAL Prefinished Red Oak or Birch	Plastic Laminate
.1 CUSTOM CABINET WORK				
See Division 11 - Stock Cabinets (Medium etc)				
Base Cabinets 35" High x 24" Deep -				
18" W	LnFt	15.45	105.00	103.00
24" W	LnFt	15.45	99.75	97.85
30" W	LnFt	14.42	94.50	92.70
36" W	LnFt	14.42	89.25	87.55
36" Wide Sink Fronts	LnFt	13.39	65.10	61.80
42"	LnFt	13.39	70.35	66.95
36" Wide Corner Cabinet	LnFt	14.42	110.25	113.30
Add for Lazy Susan	Each	20.60	105.00	103.00
Add per Drawer	Each	3.09	21.00	20.60
Upper Cabinets - 12" Deep				
30" High 18" Wide	LnFt	15.45	73.50	72.10
24" W	LnFt	15.45	68.25	66.95
30" W	LnFt	14.42	66.15	64.89
36" W	LnFt	14.42	63.00	61.80
24" High 18" Wide	LnFt	14.42	63.00	61.80
24" W	LnFt	14.42	57.75	56.65
30" W	LnFt	14.42	54.60	54.59
36" W	LnFt	13.39	54.60	53.56
15" High 18" Wide	LnFt	13.39	57.75	56.65
24" W	LnFt	13.39	55.65	54.59
30" W	LnFt	12.36	52.50	51.50
36" W	LnFt	12.36	52.50	51.50
Utility Cabinets - 84" H x 24" D				
18" Wide	LnFt	18.54	246.75	133.90
24" W	LnFt	18.54	236.25	128.75
Add - Prefinished Interior	-	-	8%	-
Deduct - Prefinished to Unfinished	-	-	12%	-
Add for Plastic Lam. Interior	-	-	-	100%
Add to Above for Special Hdwe.	Unit	7.21	-	20.60
Add - White/Red Birch/White Oak	-	-	-	20%
Add - Walnut	-	-	-	100%
China or Corner Cabinet 84"H x 36"W	Each	97.85	525.00	566.50
Oven Cabinet 84" High x 27" High	Each	92.70	420.00	442.90
Vanity Cabinets 30" High x 21" Deep				
30" Wide	LnFt	29.87	89.25	92.70
36" W	LnFt	30.90	87.15	87.55
48" W	LnFt	32.96	84.00	84.46

	UNIT	LABOR	MATERIAL
.2 COUNTER TOPS, 25" with Back Splash 4"			
Plastic Laminated 1 1/2"	LnFt	9.27	25.75
Deduct for No Back Splash	LnFt	2.06	5.15
Plastic Laminated Vanity Top	LnFt	8.24	30.90
Marble	LnFt	9.27	63.86
Artificial	LnFt	8.76	48.41
Wood Cutting Block, 1 1/2"	LnFt	8.76	61.80
Stainless Steel	LnFt	10.30	92.70
Polyester-Acrylic Solid Surface	LnFt	9.27	113.30
Granite - Artificial - 1 1/4"	LnFt	11.33	56.65
3/4"	LnFt	10.30	43.26

0603.0 MILLWORK & CUSTOM WOODWORK, Cont'd...
.3 CUSTOM DOOR FRAMES & TRIM

	Unit	Labor	Birch	Fir	Poplar	Oak	Pine	Maple
Custom Jamb & Stop								
4 5/8" x 2'6" x 3/4" x 6'8"	Each	$36	$108	$57	$56	$100	$100	-
x 2'8" x 3/4"	Each	37	110	59	57	102	100	-
x 3'0" x 3/4"	Each	38	106	63	58	103	105	-
5 1/4" x 2'6" x 3/4" x 6'8"	Each	37	117	67	58	107	102	-
x 2'8" x 3/4"	Each	38	119	69	60	110	103	-
x 3'0" x 3/4"	Each	39	122	72	62	123	110	-
6 3/4" x 2'6" x 3/4" x 6'8"	Each	39	161	-	-	163	122	-
x 2'8" x 3/4"	Each	40	165	-	-	165	123	-
x 3'0" x 3/4"	Each	41	168	-	-	168	126	-
Stock Jamb & Stop								
4 5/8" x 2'6" x 3/4" x 6'8"	Each	34	95	32	-	95	60	-
x 2'8" x 3/4"	Each	35	97	33	-	97	62	-
x 3'0" x 3/4"	Each	34	99	34	-	99	63	-
5 1/4" x 2'6" x 3/4" x 6'8"	Each	35	105	37	45	105	64	-
x 2'8" x 3/4"	Each	36	107	38	46	107	65	-
x 3'0" x 3/4"	Each	37	111	39	47	116	66	-
6 3/4" x 2'6" x 3/4" x 6'8"	Each	37	133	-	-	147	90	-
x 2'8" x 3/4"	Each	38	134	-	-	149	92	-
x 3'0" x 3/4"	Each	39	137	-	-	152	95	-
Fire Rated Jamb & Stop								
4 5/8" x 3'0" x 1 1/16"	Each	38	168	-	-	158	-	-
5 1/4" x 3'0" x 1 1/16"	Each	39	179	-	-	168	-	-
6 3/4" x 3'0" x 1 1/16"	Each	41	184	-	-	173	-	-
Trim Only -11/16" x 2 1/4"	Each	35	101	26	36	89	72	-
(2 sides) - 3/4" x 3 1/2"	Each	41	168	32	50	116	93	-

.4 MOULDINGS & TRIM-STOCK

	Unit	Labor	Birch	Fir	Poplar	Oak	Pine	Maple
Apron 7/16" x 2"	LnFt	.77	1.21	-	.53	.89	.62	-
11/16" x 2 1/2"	LnFt	.93	1.73	-	-	1.42	1.15	1.26
Astragal 1 3/4" x 2 1/4"	LnFt	.98	-	-	-	4.94	4.10	-
Base 7/16" x 2 3/4"	LnFt	.77	1.73	.60	.68	1.21	.98	1.26
9/16" x 3 1/4"	LnFt	.82	2.84	-	.95	2.21	1.25	1.37
Base Shoe 7/16" x 2 3/4"	LnFt	.77	-	-	.40	.68	.42	.68
Batten Strip 5/8" x 1 5/8"	LnFt	.67	-	-	-	1.05	.85	-
Brick Mould 1 1/4" x 3"	LnFt	1.03	-	.97	-	-	1.35	-
Casing 11/16" x 2 1/4"	LnFt	.82	1.37	.70	.63	1.00	1.00	.95
3/4" x 3 1/2"	LnFt	.88	2.36	.80	.68	1.94	1.75	-
Chair Rail 5/8" x 1 3/4"	LnFt	.70	1.73	-	1.05	1.63	1.30	-
or Dado 11/16" x 2 1/4"	LnFt	.75	2.31	-	-	2.00	1.33	-
Closet Rod 1 5/16"	LnFt	1.13	-	.80	-	1.20	-	-
Corner Bead 3/4" x 3/4"	LnFt	.67	-	-	.58	1.00	.78	1.00
1 1/8" x 1 1/8"	LnFt	.74	-	-	1.00	1.52	1.35	1.42
Cove Mould 3/4" x 7/8"	LnFt	1.03	1.10	-	.50	.79	.62	.84
Crown & Cove 9/16" x 2 3/4"	LnFt	1.08	2.73	-	.99	1.73	1.50	1.84
9/16" x 3 5/8"	LnFt	1.16	3.36	-	-	2.84	2.00	-
11/16" x 4 5/8"	LnFt	1.22	-	-	-	3.36	3.00	-
Drip Cap 11/16" x 1 3/4"	LnFt	.84	-	-	-	-	1.35	-
Half Round 1/2" x 1"	LnFt	.96	-	-	-	-	.95	-
Hand Rail 1 5/8" x 1 3/4"	LnFt	1.34	-	1.30	-	-	1.45	-
Hook Strip 5/8" x 2 1/2"	LnFt	.80	-	.80	.84	-	.95	-
Picture Mould 3/4" x 1 1/2"	LnFt	.80	.84	-	-	.95	.95	-
Quarter Round 3/4" x 3/4"	LnFt	.67	-	.36	.48	.79	.50	.79
1/2" x 1/2"	LnFt	.62	-	.27	-	-	.33	-
Sill Casement 3/4" x 1 7/8"	LnFt	1.08	-	-	-	-	1.35	-
Stool 11/16" x 2 1/2"	LnFt	1.08	3.15	1.00	1.26	2.68	1.70	-

Above prices based on 200 LnFt. Deduct for Over 200 LnFt - 20%.
Add for Walnut Trim - 100% Add for Custom Trim - 50%

0603.0   MILLWORK & CUSTOM WOODWORK, Cont'd...

		Unit	Labor	Birch	Fir	Lauan	Oak	Pine	Walnut
.5	STAIRS								
	Treads 1 1/6" x 10 1/4"	LnFt	1.65	-	-	-	4.99	3.20	-
	Risers 3/4" x 7 1/2"	LnFt	1.44	-	-	-	6.24	2.10	-
	Skirt Bds 9/16" x 9 1/2"	LnFt	1.85	-	-	-	5.72	2.30	-
	Nosings 1/8" x 3 1/2"	LnFt	1.39	-	-	-	2.86	1.55	-
.6	SHELVING								
	12" Deep	SqFt	4.12	29.00	-	-	23.00	11.00	66.00
	8"	SqFt	5.15	24.00	-	-	21.00	10.00	55.00
.7	CUSTOM PANELING	SqFt	2.58	17.00	-	-	12.00	-	20.00
.8	THRESHOLDS 3/4" x 3 1/2"								
	Standard	LnFt	6.18	-	-	-	5.20	-	-
	With Vinyl	LnFt	6.18	-	-	-	5.72	-	-

0604.0   GLUE LAMINATED (M) (Carpenters) (F.O.B. Cars)

		UNIT	LABOR	MATERIAL
.1	ARCHES AND RIGID FRAMES: Arches	MBF	546.00	1,597.00
	Rigid Frames	MBF	556.00	1,751.00
.2	BEAMS, JOISTS & PURLINS: 4" x 6", 8"& 10"	MBF	556.00	2,112.00
	6" x 6", 8" & 10"	MBF	536.00	2,009.00
	8" x 6", 8" & 10"	MBF	525.00	1,854.00
.3	DECKS:Doug.Fir - 3" x 6" (2 1/4"x 5 1/4")	SqFt	1.60	4.43
	4" x 6" (3" x 5 1/4")	SqFt	1.70	5.00
	Cedar - 3" x 6" (2 1/4"x 5 1/4")	SqFt	1.75	5.36
	4" x 6"	SqFt	1.80	5.67
	Pine - 3" x 6" (2 1/4"x 5 1/4")	SqFt	1.65	3.71

0605.0   PREFABRICATED WOOD & PLYWOOD COMPONENTS (See 0601.22, Job Fabricated)

			LABOR	MATERIAL
.1	WOOD TRUSSED RAFTERS			
	4 -12 Pitch 24" O.C. With 2' Overhang 57 lb. Loading			
	16' Span	Each	28.00	57.00
	20'	Each	30.00	60.00
	22'	Each	31.00	62.00
	24'	Each	33.00	65.00
	26'	Each	35.00	72.00
	28'	Each	39.00	84.00
	30'	Each	42.00	96.00
	32'	Each	47.00	105.00
	Add for 5-12 Pitch	Each	5%	15%
	Add for 6-12 Pitch	Each	12%	25%
	Add for Gable End Type (Flying)	Each	17.00	31.00
	Add for 16" O.C.	Each	-	10%
	Add for Hip Ends - 24' Span	Each	138.00	350.00
	Add for Energy Type	Each	-	10%
.2	WOOD FLOOR AND ROOF TRUSS JOISTS			
.21	Open Wood Web - Gusset Plate Connection			
	24" O.C. Up to 23' Span x 12" - Single Chord	LnFt	1.29	4.43
	to 24' Span x 15"	LnFt	1.34	4.53
	to 27' Span x 18"	LnFt	1.39	4.64
	to 30' Span x 21"	LnFt	1.55	4.84
	to 25' Span x 15" - Double Chord	LnFt	1.44	4.12
	to 30' Span x 18"	LnFt	1.70	4.22
	to 35' Span x 21"	LnFt	1.96	4.43
	Open Metal Web - Pin Connection			
	to 25' Span x 18"	LnFt	1.34	4.22
	to 30' Span x 22"	LnFt	1.44	4.33
	to 35' Span x 30"	LnFt	1.65	4.58
	to 40' Span x 30"	LnFt	2.06	5.67

0605.0 PREFABRICATED WOOD & PLYWOOD COMPONENTS, Cont'd...

		UNIT	LABOR	MATERIAL
.22	Plywood Web			
	24" O.C. Up to 15' Span x 9 1/2"	LnFt	1.09	2.42
	to 19' Span x 11 7/8"	LnFt	1.15	2.52
	to 21' Span x 14"	LnFt	1.29	2.88
	to 22' Span x 16"	LnFt	1.44	3.09
.3	LAMINATED VENEER STRUCTURAL BEAMS			
.31	Micro Lam (Plywood)			
	9 1/2" x 1 3/4"	LnFt	1.18	4.12
	11 7/8" x 1 3/4"	LnFt	1.24	5.15
	14" x 1 3/4"	LnFt	1.29	5.87
	16" x 1 3/4"	LnFt	1.39	6.80
	18" x 1 3/4"	LnFt	1.55	7.00
.32	Glue Lam (Dimension Lumber)			
	9" x 3 1/2" Industrial	LnFt	1.18	7.93
	12" x 3 1/2"	LnFt	1.24	9.37
	15" x 3 1/2"	LnFt	1.29	11.59
	9" x 5 1/8"	LnFt	1.18	10.92
	12" x 5 1/8"	LnFt	1.34	14.21
	15" x 5 1/8"	LnFt	1.49	17.51
	18" x 5 1/8"	LnFt	1.65	20.81
	24" x 5 1/8"	LnFt	1.96	28.22
	Add for Architectural Grade	LnFt	.21	1.03
	Add for 6 3/4"	LnFt	.21	4.38
.4	DECKS - 4' x 30' w/Plywood 2 Sides - Insulated	SqFt	.98	5.46

0606.0 WOOD TREATMENTS (Preservatives)

	UNIT	LABOR	MATERIAL
SALTS - PRESSURE (Water Borne)			
Dimension 2 x 4-2 x 6 and 2 x 8, 2 x 10-2 x 12	BdFt	-	.15
2 x 10 and 2 x 12	BdFt	-	.25
Plywood 5/8" CDX	SqFt	-	.26
SALTS - NON PRESSURE	BdFt	-	.10
CREOSOTE (Penta)	BdFt	-	.28
FIRE RETARDANTS - PRESSURE			
Dimension and Timbers	BdFt	-	.26
Plywood 5/8" CDX	SqFt	-	.26

0607.0 ROUGH HARDWARE

		UNIT	LABOR	MATERIAL
.1	NAILS, BOLTS, ETC. Job Average	SqFt & BdFt	-	.02
	Common (50# Box) 8 Penny	CWT	-	49.00
	16 Penny	CWT	-	47.00
	Finish	CWT	-	56.00
	Add for Coated	CWT	-	60%
.2	JOIST HANGERS	Each	1.29	.93
.3	BOLTS - 5/8" x 12"	Each	1.55	1.29

0608.0 EQUIPMENT (Saws, Woodworking, Drills, Etc.)

	UNIT	LABOR	MATERIAL
Average Job - As a Percentage of Labor	-	-	3%
Heavy Equipment in Division 2			
See 1-6A and 1-7A for Rental Rates			

The costs below are average, priced as total contractor/subcontractor costs with an overhead of 5% and fee of 10% included. Units include fasteners and equipment, 35% taxes and insurance on labor, and 10% General Conditions and Equipment. Also included are nails, cutting, size and lap allowances. See Example on Page 4.

0601.0 ROUGH CARPENTRY

		UNIT	COST
.1	LIGHT FRAMING AND SHEATHING		
.11	Joists and Headers - Floor Area - 16" O.C.		
	2" x 6" Joists - with Headers & Bridging	SqFt	2.00
	2" x 8" Joists	SqFt	2.51
	2" x 10" Joists	SqFt	2.82
	2" x 12" Joists	SqFt	3.49
	Add for Ceiling Joists, 2nd Floor and Above	SqFt	8%
	Add for Sloped Installation	SqFt	15%
.12	Studs, Plates and Framing - 8' Wall Height - 16" O.C.		
	2" x 3" Stud Wall - Non-Bearing (Single Top Plate)	SqFt	1.19
	2" x 4" Stud Wall - Bearing (Double Top Plate)	SqFt	1.36
	2" x 4" Stud Wall - Non-Bearing (Single Top Plate)	SqFt	1.30
	2" x 6" Stud Wall - Bearing (Double Top Plate)	SqFt	1.76
	2" x 6" Stud Wall - Non-Bearing (Single Top Plate)	SqFt	1.72
	Add for Stud Wall - 12" O.C.	SqFt	8%
	Deduct for Stud Wall - 24" O.C.	SqFt	22%
	Add for Bolted Plates or Sills	SqFt	.09
	Add for Each Foot Above 8'	SqFt	.05
	Add to Above for Fire Stops, Fillers and Nailers	SqFt	.41
	Add for Soffits and Suspended Framing	SqFt	.56
.13	Bridging - 1" x 3" Wood Diagonal	Set	2.36
	2" x 8" Solid	Each	2.56
.14	Rafters		
	2" x 4" Rafter (Incl Bracing) 3 - 12 Slope	SqFt	1.74
	4 - 12 Slope	SqFt	1.92
	5 - 12 Slope	SqFt	2.01
	6 - 12 Slope	SqFt	2.31
	Add for Hip-and-Valley Type	SqFt	.41
	Add for 1' 0" Overhang - Total Area of Roof	SqFt	.15
	2" x 6" Rafter (Incl Bracing) 3-12 Slope	SqFt	2.15
	4-12 Slope	SqFt	2.20
	5-12 Slope	SqFt	2.26
	6-12 Slope	SqFt	2.36
	Add for Hip - and - Valley Type	SqFt	.51
.15	Stairs	Each Floor	410.00
.16	Sub Floor Sheathing (Structural)		
	1" x 8" and 1" x 10" #3 Pine	SqFt	1.64
	1/2" x 4' x 8' CD Plywood - Exterior	SqFt	1.08
	5/8" x 4' x 8'	SqFt	1.23
	3/4" x 4' x 8'	SqFt	1.44
.17	Floor Sheathing (Over Sub Floor)		
	3/8" x 4' x 8' CD Plywood	SqFt	1.06
	1/2" x 4' x 8'	SqFt	1.15
	5/8" x 4' x 8'	SqFt	1.28
	1/2" x 4' x 8' Particle Board	SqFt	.99
	5/8" x 4' x 8'	SqFt	1.09
	3/4" x 4' x 8'	SqFt	1.29

0601.0   ROUGH CARPENTRY, Cont'd...

		UNIT	COST
.18	Wall Sheathing		
	1' x 8" and 1" x 10" - #3 Pine	SqFt	1.40
	3/8" x 4' x 8' CD Plywood - Exterior	SqFt	1.06
	1/2" x 4' x 8'	SqFt	1.15
	5/8" x 4' x 8'	SqFt	1.35
	3/4" x 4' x 8'	SqFt	1.58
	25/32" x 4' x 8' Fiber Board - Impregnated	SqFt	.87
	1" x 2' x 8' T&G Styrofoam	SqFt	1.11
	2" x 2' x 8'	SqFt	1.64
.19	Roof Sheathing - Flat Construction		
	1" x 6" and 1" x 8" - #3 Pine	SqFt	1.37
	1/2" x 4' x 8' CD Plywood - Exterior	SqFt	1.18
	5/8" x 4' x 8'	SqFt	1.30
	3/4" x 4' x 8'	SqFt	1.49
	Add for Sloped Roof Construction (to 5-12 slope)	SqFt	.15
	Add for Steep Sloped Construction (over 5-12 slope)	SqFt	.26
	Add to Above Sheathing (0601.16 thru 0601.19)		
	AC or AD Plywood	SqFt	.18
	10' Length Plywood	SqFt	.21
.2	HEAVY FRAMING		
.21	Columns and Beams - 16' Span Average - Floor Area	SqFt	6.77
	20' Span	SqFt	7.07
	24' Span	SqFt	8.20
.22	Deck 2" x 6" T&G - Fir Random Construction Grade	SqFt	3.90
	3" x 6" T&G - Fir Random Construction Grade	SqFt	5.13
	4" x 6"	SqFt	6.66
	2" x 6" T&G - Red Cedar	SqFt	4.92
	Add for D Grade Cedar	SqFt	1.64
	2" x 6" T&G - Panelized Fir	SqFt	3.59
.3	MISCELLANEOUS CARPENTRY		

		EACH	or	BdFt
.31	Blocking and Bucks (2" x 4" and 2" x 6")			
	Doors & Windows - Nailed to Concrete or Masonry	47.15		2.31
	Bolted to Concrete or Masonry	53.30		3.18
	Roof Edges - Nailed to Wood	-		1.79
	Bolted to Concrete (Incl. bolts)	-		2.67

		LnFt	or	BdFt
.32	Grounds and Furring			
	1" x 4" Fastened to Wood	.88		2.67
	1" x 3"	.75		2.97
	1" x 2"	.68		4.00
	1" x 3" Fastened to Concrete/Masonry - Nailed	.90		3.59
	Gun Driven	1.03		4.10
	PreClipped	1.18		4.72
	2" x 2" Suspended Framing	1.23		3.69
	Not Suspended Framing	1.03		3.08

.33	Cant Strips		
	4" x 4" Treated and Nailed	1.50	-
	Treated and Bolted (Including Bolts)	2.10	-
	6" x 6" Treated and Nailed	3.00	-
	Treated and Bolted (Including Bolts)	3.80	-

		UNIT	COST
.34	Building Papers and Sealers		
	15" Felt	SqFt	.13
	Polyethylene - 4 mil	SqFt	.12
	6 mil	SqFt	.14
	Sill Sealer	SqFt	.67

0602.0 FINISH CARPENTRY

		TO WEATHER	% ADDED	SQ FT
.1	FINISH SIDINGS AND FACING MATERIALS (Exterior)			
.11	Boards and Beveled Sidings			
	Cedar Beveled - Clear Heart 1/2" x 4"	2 3/4"	51%	4.63
	1/2" x 6"	4 3/4"	31%	3.95
	1/2" x 8"	6 3/4"	23%	3.74
	Rough Sawn 7/8" x 8"	6 3/4"	23%	3.54
	7/8" x 10"	8 3/4"	19%	3.64
	7/8" x 12"	10 3/4"	17%	3.74
	3/4" Board - Rough Sawn	-	-	3.28
	Redwood 1/2" x 4" Beveled - Clear Heart	2 3/4"	59%	4.99
	1/2" x 6"	4 3/4"	31%	4.47
	1/2" x 8"	6 3/4"	23%	4.26
	5/8" x 10"	8 3/4"	19%	3.80
	3/4" x 6"	4 3/4"	31%	4.63
	3/4" x 8"	6 3/4"	23%	4.26
	3/4" x 6" Board - Clear - 4"-6"	Varies		2.81
	3/4" Tongue & Groove - 4"-6"	Varies		4.47
	1" Rustic Beveled - 6" and 8"	Varies		4.73
	5/4" Rustic Beveled - 6" and 8"	Varies		5.36

		UNIT	COST
	Add for Metal Corners	Each	1.25
	Add for Mitering Corners	Each	1.87

			UNIT	COST
.12	Plywood Fir AC Smooth -One Side 1/4"		SqFt	1.25
	3/8"		SqFt	1.35
	1/2"		SqFt	1.72
	5/8"		SqFt	1.92
	3/4"		SqFt	2.18
	5/8" - Grooved and Rough Faced		SqFt	1.82
	Cedar - Rough Sawn 3/8"		SqFt	1.92
	5/8"		SqFt	2.18
	3/4"		SqFt	2.34
	5/8" - Grooved and Rough Faced		SqFt	2.39
	Add for Wood Batten Strips, 1" x 2" - 4' O.C.		SqFt	.31
	Add for Splines		SqFt	.33
.13	Hardboard - Paneling and Lap Siding (Primed)			
	3/8" Rough Textured Paneling -	4' x 8'	SqFt	1.72
	7/16" Grooved -	4' x 8'	SqFt	1.66
	7/16" Stucco Board Paneling		SqFt	1.82
	7/16" x 8" Lap Siding		SqFt	2.08
	7/16" x 12" Lap Siding		SqFt	1.98
	Add for Pre-Finishing		SqFt	.16
.14	Shingles			
	Wood - 16" Red Cedar - 12" to Weather	- #1	SqFt	3.22
		- #2	SqFt	2.55
		- #3	SqFt	2.44
	24" - 1/2" to 3/4" Red Cedar Hand Splits	- #1	SqFt	3.07
	24" - 3/4" to 1 1/4" Red Cedar Hand Splits	- #1	SqFt	3.17
		- #2	SqFt	2.81
		- #3	SqFt	2.60
	Add for Fire Retardant		SqFt	.42
	Add for 3/8" Backer Board		SqFt	.68
	Add for Metal Corners		Each	1.25
	Add for Ridges, Hips and Corners		LnFt	3.64
	Add for 8" to Weather		SqFt	25%

0602.0 FINISH CARPENTRY, Cont'd...

		COST		
		LN FT	or	SQ FT
.15	Facia - Pine #2 1" x 8"	2.00		3.00
	Cedar #3 1" x 8"	2.30		3.45
	Redwood - Clear 1" x 8"	2.90		4.30
	Plywood 5/8" - ACX 1" x 8"	1.60		2.40
.16	Soffit - Plywood 1/2" - ACX 1/2" Fir ACX			2.20
	Aluminum			2.75
	Plastic (Egg Crate)			2.90
.2	FINISH WALLS (Interior) (15% Added for Size & Cutting)			
.21	Boards			
	Cedar - #3 1" x 6" and 1" x 8"			3.59
	Knotty 1" x 6" and 1" x 8"			3.74
	D Grade 1" x 6" and 1" x 8"			4.42
	Aromatic 1" x 6" and 1" x 8"			4.58
	Redwood - Construction 1" x 6" and 1" x 8"			3.64
	Clear 1" x 6" and 1" x 8"			4.47
	Fir - Beaded 5/8" x 4"			3.40
	Pine - #2 1" x 6" and 1" x 8"			2.60
.22	Hardboard (Paneling) Tempered - 1/8"			1.08
	1/4"			1.22
.23	Plywood (Prefinished Paneling) 1/4" Birch - Natural			2.40
	3/4" Birch - Natural			3.35
	1/4" Birch - White			2.95
	1/4" Oak - Rotary Cut			2.40
	3/4"			3.45
	1/4" Oak - White			3.30
	1/4" Mahogany (Lauan)			1.75
	3/4"			2.60
	3/4" Mahogany (African)			4.20
	1/4" Walnut			4.00
.24	Gypsum Board (Paneling) - See 0902			

0603.0 MILLWORK & CUSTOM WOODWORK

		UNIT	COST Prefinished Red Oak or Birch	Plastic Laminate
.1	CUSTOM CABINET WORK (Red Oak or Birch)			
	Base Cabinets - Avg. 35" H x 24" D w/Drawer	LnFt	166.00	177.00
	Sink Fronts	LnFt	99.00	99.00
	Corner Cabinets	LnFt	218.00	229.00
	Add per Drawer	LnFt	42.00	42.00
	Upper Cabinets - Avg. 30" High x 12" Deep	LnFt	125.00	130.00
	Utility Cabinets - Avg. 84" High x 24" Deep	LnFt	260.00	270.00
	China or Corner Cabinets - 84" High	Each	614.00	676.00
	Oven Cabinets - 84" High x 24" Deep	Each	416.00	468.00
	Vanity Cabinets - Avg. 30" High x 21" Deep	Each	218.00	229.00
	Deduct for Prefinishing Wood	Each	12%	-
.2	COUNTER TOPS - 25"			
	Plastic Laminated with 4" Back Splash	LnFt	-	47.00
	Deduct for No Back Splash	LnFt	-	8.32
	Granite - 1 1/4" - Artificial	LnFt	-	88.00
	3/4" - Artificial	LnFt	-	68.00
	Marble	LnFt	-	94.00
	Artificial	LnFt	-	78.00
	Wood Cutting Block	LnFt	-	88.00
	Stainless Steel	LnFt	-	140.00
	Polyester-Acrylic Solid Surface	LnFt	-	156.00

(See 1103 for Stock Cabinets)

0603.0 MILLWORK & CUSTOM WOODWORK, Cont'd...

.3 CUSTOM DOOR FRAMES - Including 2 Sides Trim

	UNIT	COST
Birch	Each	250.00
Fir	Each	155.00
Poplar	Each	160.00
Oak	Each	200.00
Pine	Each	190.00
Walnut	Each	360.00

See 0802.87 for Stock Prehung Units

.4 MOULDINGS AND TRIM

	LINEAR FOOT				
	Birch	Fir	Poplar	Oak	Pine
Apron 7/16" x 2"	2.76	-	1.77	2.33	1.85
Astragal 1 3/4" x 2 1/4"	-	-	-	5.83	4.50
Base 7/16" x 2 3/4"	3.18	1.60	1.82	2.44	2.75
9/16" x 3 1/4"	4.24	-	-	3.66	3.75
Base Shoe 7/16" x 2 3/4"	-	-	1.56	2.12	2.25
Batten Strip 5/8" x 1 5/8"	-	1.24	-	2.39	1.95
Brick Mould 1 1/14" x 2"	-	2.27	-	-	2.30
Casing 11/16" x 2 1/4"	2.92	1.75	1.92	2.76	2.50
3/4" x 3 1/2"	4.03	2.27	1.98	3.92	3.40
Chair Rail 5/8" x 1 3/4"	3.18	-	2.44	3.23	2.55
Closet Rod 1 5/16"	-	2.32	-	-	2.90
Corner Bead 1 1/8" x 1 1/8"	-	2.01	2.50	3.02	2.60
Cove Moulding 3/4" x 3/4"	2.65	-	2.13	2.54	2.15
Crown Moulding 9/16" x 3 5/8"	4.88	-	3.12	3.92	3.30
11/16" x 4 5/8"	5.62	-	-	6.15	5.25
Drip Cap	-	-	-	-	2.50
Half Round 1/2" x 1"	-	-	-	-	3.00
Hand Rail 1 5/8" x 1 3/4"	-	3.45	3.80	6.04	3.40
Hook Strip 5/8" x 2 1/2"	-	1.91	2.18	-	2.25
Picture Mould 3/4" x 1 1/2"	1.91	-	-	2.80	2.50
Quarter Round 3/4" x 3/4"	-	1.08	1.40	2.23	1.70
1/2" x 1/2"	-	1.18	-	-	1.35
Sill 3/4" x 2 1/2"	-	2.27	-	-	3.00
Stool 11/16 x 2 1/2"	4.88	2.27	3.43	5.30	3.60

	UNIT	Birch	Oak	Pine
.5 STAIRS (Treads, Risers, Skirt Boards)	LnFt	33.00	25.00	18.00
.6 SHELVING (12" Deep)	SqFt	36.00	30.00	20.00
.7 CUSTOM PANELING	SqFt	22.00	17.00	14.00
.8 THRESHOLDS - 3/4" x 3 1/2"	LnFt	-	16.00	-

0604.0 GLUE LAMINATE

	UNIT	COST
Arches 80' Span	SqFt	8.20
Beams & Purlins 40' Span	SqFt	5.60
Deck Fir - 3" x 6"	SqFt	7.10
4" x 6"	SqFt	7.80
Deck Cedar - 3" x 6"	SqFt	8.40
4" x 6"	SqFt	8.80
Deck Pine - 3" x 6"	SqFt	6.50

0605.0 PREFABRICATED WOOD & PLYWOOD COMPONENTS

.1 WOOD TRUSSED RAFTERS -
24" O.C. - 4 -12 Pitch

		UNIT	COST		UNIT	COST
Span To 16'	w/ Supports	Each	113.00	or	SqFt	3.61
20'	and	Each	121.00		SqFt	3.09
22'	2' Overhang	Each	124.00		SqFt	2.78
24'		Each	129.00		SqFt	2.88
26'		Each	134.00		SqFt	2.78
28'		Each	149.00		SqFt	2.83
30'		Each	180.00		SqFt	2.88
32'		Each	201.00		SqFt	3.30
Add for 5-12 Pitch		Each	5%		SqFt	15%
Add for 6-12 Pitch		Each	12%		SqFt	25%
Add for Scissor Truss		Each	26.00		SqFt	.19
Add for Gable End 24'		Each	57.00		SqFt	.36

.2 WOOD FLOOR TRUSS JOISTS (TJI)
Open Web - Wood or Metal - 24" O.C.

		UNIT	COST		UNIT	COST
Up To 23' Span x 12"	Single	LnFt	7.42	or	SqFt	3.71
To 24' Span x 15"	Cord	LnFt	7.62		SqFt	3.81
To 27' Span x 18"		LnFt	7.83		SqFt	3.91
To 30' Span x 21"		LnFt	8.24		SqFt	4.12
To 25' Span x 15"	Double	LnFt	7.31		SqFt	3.66
To 30' Span x 18"	Cord	LnFt	7.62		SqFt	3.81
To 36' Span x 21"		LnFt	8.34		SqFt	4.17

Plywood Web - 24" O.C.

	UNIT	COST	UNIT	COST
Up To 15' x 9 1/2"	LnFt	4.64	SqFt	2.32
To 19' x 11 7/8"	LnFt	4.74	SqFt	2.37
To 21' x 14"	LnFt	5.36	SqFt	2.68
To 22' x 16"	LnFt	6.08	SqFt	3.04

.3 LAMINATED VENEER STRUCTURAL BEAMS
.31 Micro Lam - Plywood - 24" O.C.

	UNIT	COST		UNIT	COST
9 1/2" x 1 3/4"	LnFt	6.90	or	SqFt	3.45
11 7/8" x 1 3/4"	LnFt	8.03		SqFt	4.02
14" x 1 3/4"	LnFt	9.27		SqFt	4.64
16" x 1 3/4"	LnFt	10.30		SqFt	5.15

.32 Glue Lam- Dimension Lumber- 24" O.C.

	UNIT	COST	UNIT	COST
9" x 3 1/2"	LnFt	11.54	SqFt	5.77
12" x 3 1/2"	LnFt	13.60	SqFt	6.80
15" x 3 1/2"	LnFt	16.48	SqFt	8.24
9" x 5 1/2"	LnFt	15.35	SqFt	7.67
12" x 5 1/2"	LnFt	19.36	SqFt	9.68
15" x 5 1/2"	LnFt	25.13	SqFt	12.57
18" x 5 1/2"	LnFt	28.63	SqFt	14.32
Add for Architectural Grade	LnFt	25%		
Add for 6 3/4"	LnFt	40%		

.4 DECKS 4' x 30' w/ Plywood - Insulated

	UNIT	COST
	SqFt	8.00

0607.0 ROUGH HARDWARE

		UNIT	COST
.1	NAILS	SqFt or BdFt	.02
.2	JOIST HANGERS	Each	2.70
.3	BOLTS - 5/8" x 12"	Each	3.20

		PAGE
0701.0	**WATERPROOFING (07100)**	**7-2**
.1	MEMBRANE	7-2
.2	HYDROLITHIC	7-2
.3	ELASTOMERIC (Liquid)	7-2
.4	METALLIC OXIDE	7-2
.5	VINYL PLASTIC	7-2
.6	BENTONITE	7-2
0702.0	**DAMPPROOFING (07150)**	**7-2**
.1	BITUMINOUS	7-2
.2	CEMENTITOUS	7-2
.3	SILICONE	7-2
0703.0	**BUILDING INSULATION (07210)**	**7-2**
.1	FLEXIBLE (Fibrous Batts & Rolls)	7-2
.2	RIGID	7-3
.3	LOOSE	7-3
.4	FOAMED	7-3
.5	SPRAYED	7-3
.6	ALUMINUM	7-3
.7	VINYL FACED FIBERGLASS	7-3
0704.0	**ROOF & DECK INSULATION (07240)**	**7-4**
0705.0	**MEMBRANE ROOFING (07500)**	**7-4**
.1	BUILT UP	7-4
0706.0	**SINGLE PLY ROOFING**	**7-4**
0707.0	**TRAFFIC ROOF COATING (07570)**	**7-5**
.1	PEDESTRIAN	7-5
.2	VEHICULAR	7-5
0708.0	**SHINGLE & TILE ROOFING (07300)**	**7-6**
.1	ASPHALT SHINGLES	7-6
.2	FIBERGLASS SHINGLES	7-6
.3	WOOD SHINGLES	7-6
.4	METAL SHINGLES	7-6
.5	SLATE ROOFING	7-6
.6	TILE ROOFING	7-6
.61	Clay	7-6
.62	Concrete	7-6
.63	Steel	7-6
.7	CORRUGATED ROOFING	7-6
0709.0	**SHEET METAL & FLASHING (07600)**	**7-6**
.1	DOWNSPOUTS, GUTTERS & FITTINGS	7-6
.2	FACINGS, GRAVEL STOPS & TRIM	7-7
.3	FLASHINGS	7-7
.4	SHEET METAL ROOFING	7-7
.5	EXPANSION JOINTS	7-7
0710.0	**ROOF ACCESSORIES (07700)**	**7-8**
.1	SKYLIGHTS	7-8
.2	SKY DOMES	7-8
.3	GRAVITY VENTILATORS	7-8
.4	SMOKE VENTILATORS	7-8
.5	PREFAB CURBS & EXPANSION JOINTS	7-8
.6	ROOF LINE LOUVRES	7-8
0711.0	**SEALANTS (07900)**	**7-8**
.1	CAULKING	7-8
.2	JOINT FILLERS & GASKETS	7-8
7A	**QUICK ESTIMATING**	**7A-9 & 7A-10**

		UNIT	LABOR	MATERIAL
0701.0	**WATERPROOFING (L&M) (Roofers)**			
.1	MEMBRANE			
	1 Ply Membrane - Felt 15#	SqFt	.45	.32
	Fabric and Felt	SqFt	.64	.36
	2 Ply Membrane - Felt 15#	SqFt	.78	.52
	Fabric and Felt	SqFt	.90	.56
.2	HYDROLITHIC	SqFt	.79	.79
.3	ELASTOMERIC- Rubberized Asphalt .060M & Poly	SqFt	.79	.53
	Liquid	SqFt	.80	.66
.4	METALLIC OXIDE- 3 Coat	SqFt	.74	1.68
	3 Coat with Sand & Cement Cover	SqFt	.95	1.85
.5	VINYL PLASTIC	SqFt	.48	1.00
.6	BENTONITE - 3/8"	SqFt	.64	.59
0702.0	**DAMPPROOFING (Roofers or Laborers)**			
.1	BITUMINOUS (Asphalt)			
.11	Trowel Mastics			
	Asphalt 1/16" (12 Sq.Ft. Gal.)	SqFt	.28	.21
	1/8" (20 Sq.Ft. Gal.)	SqFt	.33	.26
.12	Spray or Brush Liquid			
	Asphalt - 1 Coat - Spray	SqFt	.19	.13
	Spray	SqFt	.25	.18
	Brush	SqFt	.29	.20
.13	Hot Mop			
	Asphalt - 1 Coat - Including Primer	SqFt	.46	.34
	Fibrous Asphalt - 1 Coat	SqFt	.52	.37
.14	Trowel Mastics & Pre-formed Vapor Barriers			
	Mastic Fabric Mastic	SqFt	.48	.71
	and Polyvinyl Chloride	SqFt	.46	.59
.2	CEMENTITIOUS - 1 Coat - 1/2"	SqFt	.28	.14
	2 Coat - 1"	SqFt	.38	.18
.3	SILICONE (Masonry) - 1 Coat	SqFt	.21	.15
	2 Coat	SqFt	.28	.25
	Add for Scaffold to Above Sections	SqFt	.20	.10
0703.0	**BUILDING INSULATION (S) (Carpenters)**			
.1	FLEXIBLE (Batts and Rolls) - Friction			
	Fiberglass			
	2 1/4" R 7.4	SqFt	.22	.24
	3 1/2" R 11.0	SqFt	.23	.26
	3 5/8" R 13	SqFt.	.24	.33
	6" R 19.0	SqFt	.26	.39
	9" R 30.0	SqFt	.30	.66
	Add for Paper Faced Batts	SqFt	.02	.03
	Add for Polyethylene Barrier 2 Mil	SqFt	.04	.04
	Add for Staggered Stud Installation	SqFt	.08	.03

0703.0 BUILDING INSULATION, Cont'd...

	UNIT	LABOR	MATERIAL
.2 RIGID			
Fiberglass			
1" R 4.35 3# Density	SqFt	.27	.33
1 1/2" R 6.52 3# Density	SqFt	.29	.48
2" R 8.70 3# Density	SqFt	.31	.63
Add for 4.2# Density per Inch	SqFt	-	.07
Add for 6.0# Density per Inch	SqFt	-	.25
Add for Kraft Paper Backed	SqFt	-	.08
Add for Aluminum Foil Backed	SqFt	-	.35
Add for Fireproof Type	SqFt	-	10%
Expanded Styrene			
Molded, White (Bead Board)			
1" x 16" x 8' R 5.40	SqFt	.27	.19
1" x 24" x 8', T&G R 5.40	SqFt	.30	.26
1 1/2" R 6.52	SqFt	.29	.28
2" R 7.69	SqFt	.31	.36
2" x 24" x 8', T&G R 7.69	SqFt	.34	.52
Extruded, Blue			
1" x 24" x 8' R 5.40	SqFt	.27	.38
T&G	SqFt	.30	.48
1 1/2" R 6.52	SqFt	.29	.48
2" R 7.69	SqFt	.31	.50
2" x 24" x 8', T&G R 7.69	SqFt	.34	.75
Perlite			
1" R 2.78	SqFt	.27	.43
2" R 5.56	SqFt	.33	.75
Urethane			
1" R 6.67	SqFt	.30	.78
2" R 13.34	SqFt	.35	1.45
.3 LOOSE			
Expanded Styrene 1" R 3.80	SqFt	.21	.21
Fiberglass 1" R 2.20	SqFt	.22	.22
Rockwool 1" R 2.90	SqFt	.23	.22
Cellulose 1" R 3.70	SqFt	.21	.23
.4 FOAMED			
Urethane per Inch or Bd.Ft.	SqFt	.43	.80
.5 SPRAYED			
Fibered Cellulose per Inch or Bd.Ft.	SqFt	.38	.42
Polystyrene per Inch or Bd.Ft.	SqFt	.45	.50
Urethane per Inch or Bd.Ft.	SqFt	.50	.90
.6 ALUMINUM PAPER	SqFt	.14	.14
.7 VINYL FACED FIBERGLASS			
2 5/8" x 63 1/2" x 96" R 10.0	SqFt	.35	.85
Add to All Above for Ceiling Work	SqFt	.06	-
Add to All Above for Clip On	SqFt	.07	.02
Add to All Above for Scaffold	SqFt	.18	.10

See Division 3 for Cementitious Poured and Rigid Insulation
See Division 4 for Core and Cavity Filled Masonry
See Division 7 for Roof Insulation
See Division 15 for Mechanical Insulations

		UNIT	COST
0704.0	**ROOF AND DECK INSULATION (L&M) (Roofers)**		
.1	RIGID - NON RATED		
	Fiber Board 1/2" R 1.39	Sq	46.00
	3/4" R 2.09	Sq	56.00
	1" R 2.78	Sq	65.00
	Fiberglass 1" R 3.70	Sq	75.00
	1 1/2" R 2.09	Sq	95.00
	Perlite 1/2" R 1.39	Sq	45.00
	3/4" R 2.09	Sq	56.00
	1" R 2.78	Sq	66.00
	2" R 5.26	Sq	98.00
	Expanded Polystrene 1" R 4.35	Sq	60.00
	1 1/2" R 6.52	Sq	70.00
	2" R 7.69	Sq	88.00
.2	RIGID - FIRE RATED		
	Isocyanurate 1" R 6.30 Glass Faced	Sq	100.00
	1 1/5" R 7.60	Sq	105.00
	1 1/2" R 10.00	Sq	110.00
	1 3/4" R 12.00	Sq	120.00
	2" R 14.00	Sq	125.00
	Phenolic Foam 1" R 7.60	Sq	83.00
	1 1/2" R 12.50	Sq	93.00
	2" R 16.60	Sq	115.00
	Add for 5/8" Sheet Rock over Steel Deck	Sq	13.00
	Add for Fiber or Mineral Cants	LnFt	1.40
	Add for Felt Faces	Sq	9.00
	Add for Composite Insulations	Sq	25%
.3	SPRAYED - 2" Polystyrene	Sq	225.00
0705.0	**MEMBRANE ROOFING (L&M) (Roofers)**		
	BUILT-UP ROOFING (50 SQUARES OR MORE)		
	Asphalt & Gravel - 15# Glass/Organic Felts 3-Ply	Sq	130.00
	4 - Ply	Sq	140.00
	5 - Ply	Sq	155.00
	Add for Pitch and Gravel	Sq	40.00
	Add for Curbed Openings	Each	115.00
	Add for Round Vent Openings	Each	105.00
	Add for Roof Skids	LnFt	23.00
	Add for Sloped Roofs	Sq	30.00
	Add for Applications on Wood	Sq	10.00
	Add for Marble Chips	Sq	45.00
	Add for Barriers - 15" Felt	Sq	9.00
	Fire Resistant	Sq	20.00
	Add for Walkways - 1"	SqFt	4.25
	Add for Bond	Sq	10.00
	Add for Roof Openings Cut in Existing	Each	380.00

See Quick Estimating 7A-10 for Combined Roofing, Insulation and Sheet Metal

0706.0 SINGLE-PLY ROOFING No Insulation or Flashing Included

	UNIT	Gravel Ballast	Application Mechanical Fastened	Adhesive Fastened
Butylene 100 mil	Sq	165.00	180.00	185.00
EPDM 50 mil	Sq	130.00	155.00	180.00
Neoprene 60 mil	Sq	240.00	290.00	295.00
P.V.C. 50 mil	Sq	120.00	175.00	195.00
Add per Opening	Opng	115.00	-	-

0707.0 TRAFFIC ROOF COATINGS (L&M)

		UNIT	LABOR	MATERIAL
.1	PEDESTRIAN TRAFFIC TYPE	SqFt	-	5.75
.2	VEHICULAR TRAFFIC TYPE			
	Rubberized Coating with 3/4" Asphalt Topping	SqFt	-	3.85
	Polyurethane with Non-Slip Aggregates			
	on 20 mil silicone rubber	SqFt	-	3.45
	Elastomeric with Non-Skid - 2 Coat	SqFt	-	2.20
	3 Coat	SqFt	-	2.90

0708.0 SHINGLE, TILE and CORRUGATED ROOFING

		UNIT	LABOR	MATERIAL
.1	ASPHALT SHINGLES (S) (Carpenters) 4-12 Pitch			
	Square Butt 235#- Seal Down	SqFt	.38	.32
	300# - Laminated	SqFt	.43	.54
	325# - Fire Resistant	SqFt	.46	.52
	Timberline 300# - 25 Year	SqFt	.52	.44
	300# - 30 Year	SqFt	.54	.60
	Tab Lock 340#	SqFt	.52	.80
	Roll 90# - 100 SqFt to Roll	SqFt	.24	.16
	55# - 100 SqFt to Roll	SqFt	.20	.14
	Base Starter - 240 SqFt to Roll	SqFt	.20	.20
	Ice & Water Starter - 108 SqFt to Roll	SqFt	.25	.90
	Add for Boston Ridge Shingle	LnFt	1.00	.40
	Add for 15# Felt Underlayment	SqFt	.06	.04
	Add for Pitches 6-12 & Up - Ea Pitch Increase	SqFt	.06	-
	Add for Removal and Haul - Away (2 Layers)	SqFt	.27	.12
	Add for Chimneys, Skylights & Bay Windows	Each	40.00	-
.2	FIBERGLASS SHINGLES - 215# - Seal Down	SqFt	.38	.32
	250# - Seal Down	SqFt	.43	.50
.3	WOOD			
	Sawed			
	Red Cedar - 16" x 5" to Weather - #1 Grade	SqFt	.70	1.50
	#2 Grade	SqFt	.68	1.20
	#3 Grade	SqFt	.64	.75
	Hand Splits			
	Red Cedar - 1/2" x 3/4", 24" x 10" to Weather	SqFt	.87	1.20
	3/4"x1 1/4", 24" x 10" to Weather	SqFt	.90	1.50
	Fancy Butt - 7" to Weather	SqFt	1.00	3.20
	Add for Fire Retardant	SqFt	-	.40
	Add for Pitches over 5-12, Each Pitch Increase	SqFt	.06	-
.4	METAL			
	Aluminum - Mill .020	SqFt	.74	1.50
	.030	SqFt	.75	1.65
	Anodized .020	SqFt	.74	1.90
	.030	SqFt	.75	2.15
	Steel - Enameled - Colored	SqFt	.79	2.60
	Galvanized - Colored	SqFt	.79	2.00

0708.0 SHINGLE, TILE AND CORRUGATED ROOFING, Cont'd...

.6 SLATE SHINGLES (Incl. Underlayment) (L&M) (Roofers)

Vermont and Pennsylvania	UNIT	COST
3/16" - 6 1/2" to 8 1/2" to Weather		
Weathered Green or Black	Sq	980.00
Gray	Sq	990.00
Fading Purple	Sq	1,020.00
Unfading Purple	Sq	1,240.00
Add for Variegated Colors & Sizes - 1/4" to 1/2"	Sq	330.00
3/8" to 1"	Sq	640.00
Add for Hips and Valleys	LnFt	50.00
Add for Pitches over 6-12	Sq	70.00

.7 TILE (L&M) (Roofers)

	UNIT	COST
.71 Clay, Interlocking Shingle Type - Designed	Sq	970.00
Early American	Sq	1,010.00
Williamsburg	Sq	1,070.00
Mormon	Sq	1,560.00
Architectural Pattern Shingle Type - Provincial	Sq	930.00
Georgian	Sq	930.00
Colonial	Sq	930.00
Spanish, Red	Sq	660.00
Mission	Sq	1,090.00
.72 Concrete - Plain - 10" to Weather - 9" Wide	Sq	590.00
Colored	Sq	660.00
Add for Staggered Installation	Sq	215.00
.73 Steel - 16 Ga - Ceramic Coated	Sq	400.00
.74 Aluminum .019	Sq	475.00

.8 CORRUGATED PANELS

	UNIT	COST
.81 Concrete - Plain	Sq	3.40
.82 Vinyl - 120	Sq	5.35
.83 Fiberglass - 8 oz	Sq	3.60
.84 Aluminum .024 - Plain	Sq	3.05
.024 - Painted	Sq	3.70

See Division 5 for Corrugated Metal Panels

0709.0 SHEET METAL WORK (L&M)

(Sheet Metal Workers)

	UNIT	COST Galvanized 25 Ga	Aluminum .032"	Copper 16 oz	Enameled
.1 DOWNSPOUTS, GUTTERS AND FITTINGS					
Downspouts - 3" x 2"	LnFt	2.35	2.35	6.40	2.90
4" x 3"	LnFt	2.90	3.10	9.10	3.90
5" x 4"	LnFt	4.00	4.10	9.50	5.00
Round 3"	LnFt	2.25	2.75	5.90	2.90
4"	LnFt	2.65	3.30	7.00	3.70
5"	LnFt	3.35	4.00	7.80	4.45
Chute Type 4" x 6"	LnFt	3.85	4.50	-	5.35
5" x 7"	LnFt	4.40	5.05	-	6.20
Gutters - Box or Half Round 4"	LnFt	3.10	4.00	6.35	4.20
5"	LnFt	3.60	4.10	7.90	4.65
6"	LnFt	4.40	4.70	8.10	5.70
Add for Guards	LnFt	-	1.00	-	-
Fittings (Elbows, Corners, etc.)	LnFt	9.30	8.80	11.00	11.00

See Quick Estimating 7A-10 for Combined Roofing, Insulation and Sheet Metal.

0709.0 SHEET METAL WORK, Cont'd...

.2 FASCIAS, GRAVEL STOPS & TRIM

	UNIT	Galvanized 25 Ga	Aluminum .032"	Copper 16 oz	Enameled	PVC
Fascias - 6"	LnFt	4.40	6.80	9.80	5.40	5.90
8"	LnFt	4.70	7.60	10.80	5.70	6.50
Copings - 12"	LnFt	5.60	9.00	11.30	6.75	-
Scuppers	Each	105.00	105.00	125.00	110.00	-
Gravel Stops - 4"	LnFt	4.30	6.25	7.40	5.50	-
6"	LnFt	4.00	7.10	9.25	5.70	6.05
8"	LnFt	5.20	7.50	10.50	6.30	6.50
Add for Durodonic Finish	LnFt		25%			

.3 FLASHINGS

	UNIT	COST
Galvanized - 25 ga	SqFt	4.30
Aluminum - .019	SqFt	4.00
.032	SqFt	4.30
Copper - 16 oz	SqFt	7.00
20 oz	SqFt	7.20
Asphalted Fabric - Plain	SqFt	2.15
Copper Backed	SqFt	3.50
Aluminum Back	SqFt	3.10
Vinyl Fabric - .020	SqFt	1.15
.030	SqFt	1.30
Lead - 2.5#	SqFt	6.45
Rubber - 1/16"	SqFt	2.60
PVC - 24 mil	SqFt	2.90

.4 SHEET METAL ROOFING - Including Underlayment

	UNIT	COST
Copper - Standing Seam - 16 oz	Sq	910.00
20 oz	Sq	1,030.00
Batten Seam 16 oz	Sq	920.00
20 oz	Sq	1,050.00
Flat Lock 16 oz	Sq	850.00
20 oz	Sq	970.00
Stainless Steel - Batten Seam 28 ga	Sq	930.00
Standing Seam 28 ga	Sq	1,030.00
Flat Seam 28 ga	Sq	880.00
Monel - Batten Seam .018"	Sq	1,050.00
Standing Seam .018"	Sq	1,080.00
Flat Seam .108"	Sq	1,000.00
Lead - Copper Coated - Batten Seam 3 lb	Sq	985.00
Flat Seam 3 lb	Sq	930.00
Aluminum - Clad	Sq	760.00
Galvanized - Colored .020"	Sq	825.00
See Division 5 for Corrugated Type		

.5 EXPANSION JOINTS - 2 1/2"

	UNIT	COST
Polyethylene with galvanized	LnFt	6.20
Galvanized 25 ga	LnFt	7.25
Aluminum .032	LnFt	8.25
Copper 16 oz	LnFt	20.00
Stainless Steel	LnFt	10.90
Neoprene	LnFt	7.95
Butyl	LnFt	6.40
Add for 3 1/2"	LnFt	10%

0710.0	ROOF AND SOFFITT ACCESSORIES (L&M) (Roofers)	UNIT	COST	or	UNIT	COST
.1	SKYLIGHTS					
	Example: 4' - 0" x 4' - 0"	SqFt	35.00		Each	560.00
	8' - 0" x 8' - 0"	SqFt	22.00		Each	1,400.00
.2	SKY DOMES (Plastic - Clear or White)					
	20" x 20" Single				Each	155.00
	Double				Each	185.00
	24" x 24" Single				Each	190.00
	Double				Each	215.00
	48" x 48" Single				Each	450.00
	Double				Each	520.00
	72" x 72" Single				Each	760.00
	Double				Each	930.00
	96" x 96" Double				Each	1,500.00
	Add for Bronze Top Colors				-	10%
	Add for Electric Operated Type				Each	435.00
	Add for Pyramid Type				Each	100%
	Add for Hip Type				Each	50%
.3	GRAVITY VENTILATORS					
	Diameter Aluminum or Galvanized 12"				Each	100.00
	24"				Each	240.00
.4	SMOKE VENTILATORS					
	Example: 48" x 96" - Aluminum				Each	2,000.00
	48" x 48" - Steel				Each	1,900.00
.5	PREFAB CURBS & EXPANSION JOINTS - 3" Fibered				LnFt	1.55
	4" Fibered				LnFt	1.60
.6	ROOF LINE LOUVRES				LnFt	1.50
.7	TURBINE VENTILATORS - 12"				Each	300.00
	24"				Each	310.00
.8	CUPOLAS				Each	225.00
.9	ROOF VENTS - P.V.C.				Each	57.00
	Aluminum				Each	52.00
	See 1012 for Roof Hatches					
0711.0	SEALANTS (L&M) (Carpenters)					
.1	CAULKING - 1/2" x 1/2" Joints					
	Windows and Doors - 2-Part Proprietary (Oil Base)				LnFt	1.65
	Silicone Rubber				LnFt	1.90
	Polysulfide (Thiakol)				LnFt	2.05
	Acrylic				LnFt	2.00
	Polyurethane				LnFt	2.00
	Control Joints - 2 - Part Proprietary				LnFt	1.80
	Silicone Rubber				LnFt	2.15
	Polysulfide				LnFt	2.95
	Acrylic				LnFt	2.15
	Polyurethane				LnFt	2.15
	Stone Pointing -Silicone Rubber				LnFt	2.45
	Polysulfide				LnFt	2.70
	Acrylic				LnFt	2.15
	Add for Swing Stage Work				LnFt	.40
	Add for 3/4" x 3/4" Joints				LnFt	.80
	Deduct for 1/4" x 1/4" Joints				LnFt	.25
.2	GASKETS AND JOINT FILLERS					
	Gaskets - 1/4" Polyvinyl x 6"				LnFt	2.00
	1/2" Polyvinyl x 6"				LnFt	2.45
	1/4" Neoprene x 6"				LnFt	2.60
	1/2" Neoprene x 6"				LnFt	3.40
	Joint Fillers -1/2"				LnFt	.30
	3/4"				LnFt	.34
	1"				LnFt	.38

	SQ.FT. COST
0701.0 WATERPROOFING	
1-Ply Membrane Felt 15#	1.20
2-Ply Membrane Felt 15#	1.85
Hydrolithic	2.15
Elastomeric Rubberized Asphalt with Poly Sheet	1.95
Liquid	2.15
Metallic Oxide 3-Coat	3.20
Vinyl Plastic	2.20
Bentonite 3/8" - Trowel	1.55
5/8" - Panels	1.75
0702.0 DAMPPROOFING	
Asphalt Trowel Mastic 1/16"	.72
1/8"	.86
Spray Liquid 1-Coat	.45
2-Coat	.63
Brush Liquid 2-Coat	.70
Hot Mop 1-Coat and Primer	1.20
1 Fibrous Asphalt	1.30
Cementitious Per Coat 1/2" Coat	.62
Silicone 1-Coat	.45
2-Coat	.68
Add for Scaffold and Lift Operations	.40
0703.0 BUILDING INSULATION	
FLEXIBLE	
Fiberglass 2 1/4" R 7.40	.64
3 1/2" R 11.00	.68
3 5/8" R 13.00	.76
6" R 19.00	.87
9" R 30.00	1.20
12" R 38.00	1.45
Rock Wool 3 1/2" R 11.00	.64
4" R 13.00	.71
6" to 7" R 19.00	.87
9" R 30.00	1.20
Add for Ceiling Work	.08
Add for Paper Faced	.08
Add for Polystyrene Barrier (2m)	.08
Add for Scaffold Work	.40
RIGID	
Fiberglass 1" R 4.35 3# Density	.81
1 1/2" R 6.52 3# Density	1.06
2" R 8.70 3# Density	1.30
Styrene, Molded 1" R 4.35	.64
1 1/2" R 6.52	.74
2" R 7.69	.88
2" T&G R 7.69	.98
Styrene, Extruded 1" R 5.40	.93
1 1/2" R 6.52	1.05
2" R 7.69	1.30
2" T&G R 7.69	1.46
Perlite 1" R 2.78	.95
2" R 5.56	1.46
Urethane 1" R 6.67	1.15
2" R 13.34	2.15
Add for Glued Applications	.08

	SQ.FT.COST
0703.0 BUILDING INSULATION, Cont'd...	
LOOSE 1" Styrene R 3.8	.58
1" Fiberglass R 2.2	.54
1" Rock Wool R 2.9	.54
1" Cellulose R 3.7	.50
FOAMED Urethane Per Inch	1.50
SPRAYED Cellulose Per Inch	1.05
Polystyrene	1.25
Urethane	1.90
ALUMINUM PAPER	1.55
VINYL FACED FIBERGLASS	1.70
0704.0 ROOF & DECK INSULATION - COMBINED WITH 0709.0 BELOW	
0705.0 MEMBRANE ROOFING - COMBINED WITH 0709.0 BELOW	
0708.0 SHINGLE ROOFING	
ASPHALT SHINGLES 235# Seal Down	1.00
300# Laminated	1.30
325" Fire Resistant	1.40
Timberline	1.60
340# Tab Lock	1.70
ASPHALT ROLL 90#	.50
55#	.40
FIBERGLASS SHINGLES 215#	1.00
250#	1.30
RED CEDAR 16" x 5" to Weather #1 Grade	2.80
#2 Grade	2.20
#3 Grade	1.90
24" x 10" to Weather - Hand Splits - 1/2" x 3/4" #2	2.50
3/4" x 1 1/4" #2	2.90
Add for Fire Retardant	.50
METAL Aluminum 020 Mill	2.65
020 Anodized	3.20
Steel, Enameled Colored	4.00
Galvanized Colored	3.25
Add to Above for 15# Felt Underlayment	
Add for Pitches Over 5-12 each Pitch Increase	.06
Add for Chimneys, Skylights and Bay Windows - Each	60.00

0704.0/ 0705.0/ 0709.0 - COMBINED ROOFING INSULATION and SHEET METAL	
MEMBRANE	SQ. COST
3-Ply Asphalt & Gravel - Fiberboard & Perlite - R 10.0	370.00
R 16.6	420.00
4-Ply - R 10.0	425.00
R 16.6	445.00
5-Ply - R 10.0	485.00
R 16.6	525.00
Add for Sheet Rock over Steel Deck - 5/8"	100.00
Add for Fiberglass Insulation	32.00
Add for Pitch and Gravel	50.00
Add for Sloped Roofs	40.00
Add for Thermal Barrier	32.00
Add for Upside Down Roofing System	92.00
SINGLE-PLY - 60M Butylene Roofing & Gravel Ballast - R 10.0	340.00
Mech. Fastened - R 10.0	360.00
PVC & EPDM Roofing & Gravel Ballast - R 10.0	320.00
Mech. Fastened - R 10.0	355.00
Blocking and Cants Not Included	

		PAGE
0801.0	**HOLLOW METAL DOORS & FRAMES (CSI 08100)**	**8-3**
.1	CUSTOM	8-3
.11	Frames	8-3
.12	Doors	8-3
.2	STOCK	8-3
.21	Frames	8-3
.22	Doors	8-3
0802.0	**WOOD DOORS (CSI 08200)**	**8-4**
.1	FLUSH DOORS	8-4
.2	PANEL DOORS	8-4
.3	LOUVERED DOORS	8-5
.4	BI-FOLD DOORS - FLUSH	8-5
.5	CAFE DOORS	8-5
.6	FRENCH DOORS	8-5
.7	DUTCH DOORS	8-5
.8	PREHUNG DOOR UNITS	8-5
0803.0	**SPECIAL DOORS (CSI 08300)**	**8-6**
.1	BLAST OR NUCLEAR RESISTANT (CSI 08315)	8-6
.2	CLOSET - BI-FOLDING (CSI 08350)	8-6
.21	Wood	8-6
.22	Metal	8-6
.23	Leaded Metal	8-6
.3	FLEXIBLE DOORS (CSI 08350)	8-6
.4	GLASS - ALL (CSI 08340)	8-6
.5	HANGAR & INDUSTRIAL (CSI 08370)	8-6
.6	METAL COVERED FIRE & INDUSTRIAL SLIDING DOORS (CSI 08300)	8-6
.7	OVERHEAD DOORS - COMMERCIAL & RESIDENTIAL (CSI 08360)	8-7
.8	PLASTIC LAMINATE FACED (CSI 08220)	8-7
.9	REVOLVING DOORS (CSI 08470)	8-7
.10	ROLLING DOORS & GRILLES (CSI 08330 & 08350)	8-8
.11	SHOWER DOORS (CSI 10820)	8-8
.12	SLIDING OR PATIO DOORS (CSI 08640)	8-8
.13	SOUND REDUCTION (CSI 08385)	8-8
.14	TRAFFIC (CSI 08380)	8-8
.15	SPECIAL MADE & ENGINEERED INDUSTRIAL DOORS (CSI 08370)	8-8
.16	VAULT DOORS (CSI 11030)	8-8
0804.0	**ENTRANCE DOORS, FRAMES & STORE FRONT CONSTRUCTION (CSI 08400)**	**8-9**
.1	ALUMINUM	8-9
.11	Stock	8-9
.12	Custom	8-9
.2	BRONZE	8-9
.3	STAINLESS STEEL	8-9
0805.0	**METAL WINDOWS (CSI 08500)**	**8-9**
.1	ALUMINUM WINDOWS	8-9
.2	ALUMINUM SASH	8-9
.3	STEEL WINDOWS	8-9
.4	STEEL SASH	8-9
.5	AWNING	8-9
0806.0	**WOOD & VINYL CLAD WINDOWS (CSI 08610)**	**8-10**
.1	BASEMENT OR UTILITY	8-10
.2	CASEMENT OR AWNING	8-10
.3	DOUBLE HUNG	8-10
.4	GLIDER WITH SCREEN	8-10
.5	PICTURE WINDOW WITH SCREENS	8-11
.6	CASEMENT ANGLE BAY WINDOWS	8-11
.7	CASEMENT BOW WINDOWS	8-11
.8	90° CASEMENT BOX BAY WINDOWS	8-11
.9	SKY OR ROOF WINDOWS	8-11
.10	CIRCLE TOPS AND ROUND	8-11

		PAGE
0807.0	**SPECIAL WINDOWS (CSI 08650)**	**8-12**
.1	LIGHT-PROOF WINDOWS (CSI 08667)	8-12
.2	PASS WINDOWS (CSI 08665)	8-12
.3	DETENTION WINDOWS (CSI 08660)	8-12
.4	VENETIAN BLIND WINDOWS	8-12
.5	SOUND CONTROL WINDOWS (CSI 08653)	8-12
0808.0	**DOOR AND WINDOW ACCESSORIES**	**8-12**
.1	STORMS AND SCREENS (CSI 08670)	8-12
.2	DETENTION SCREENS (CSI 08660)	8-12
.3	DOOR OPENING ASSEMBLIES (CSI 08720)	8-12
.4	SHUTTERS - FOLDING (CSI 08668) & ROLL-UP (CSI 08664)	8-12
0809.0	**CURTAIN WALL SYSTEMS (CSI 08900)**	**8-12**
.1	STRUCTURAL OR VERTICAL TYPE	8-12
.2	PANEL WALL OR HORIZONTAL TYPE	8-12
.3	ARCHITECTURAL PANELS	8-12
0810.0	**FINISH HARDWARE (CSI 08710)**	**8-13**
.1	BUTTS	8-13
.2	CATCHES - ROLLER	8-13
.3	CLOSURES	8-13
.4	DEAD BOLT LOCKS	8-13
.5	EXIT DEVICES	8-13
.6	HINGES, SPRING	8-13
.7	LATCHSETS	8-13
.8	LOCKSETS	8-13
.9	PLATES	8-13
.10	STOPS AND HOLDERS	8-13
0811.0	**WEATHER-STRIPPING & THRESHOLDS (CSI 08730)**	**8-13**
.1	ASTRAGALS	8-13
.2	DOORS	8-13
.3	SWEEPS	8-13
.4	THRESHOLDS	8-13
.5	WINDOWS	8-13
0812.0	**GLASS AND GLAZING (CSI 08810)**	**8-14**
.1	BEVELED GLASS	8-14
.2	COATED	8-14
.3	HEAT ABSORBING, et al	8-14
.4	INSULATED GLASS	8-14
.5	LAMINATED (SAFETY AND SOUND)	8-14
.6	MIRROR GLASS	8-14
.7	PATTERNED OR ROUGH (OBSCURE)	8-14
.8	PLASTIC GLASS (SAFETY)	8-14
.9	PLATE - CLEAR	8-14
.10	SANDBLASTED	8-14
.11	SHEET GLASS	8-14
.12	STAINED GLASS	8-14
.13	STRUCTURAL GLASS	8-14
.14	TEMPERED (SAFETY PLATE)	8-14
.15	TINTED PLATE - LIGHT REFLECTIVE	8-14
.16	WIRED (SAFETY)	8-14
8A	**QUICK ESTIMATING SECTIONS**	**8A-15 thru 8A-20**

DIVISION #8 - DOORS, WINDOWS & GLASS

0801.0   HOLLOW METAL (M) (CARPENTERS)

	UNIT	LABOR	MATERIAL
.1 CUSTOM			
.11 Frames (16 Gauge)			
2'6" x 6'8" x 4 3/4"	Each	44	130
2'8" x 6'8" x 4 3/4"	Each	45	140
3'0" x 6'8" x 4 3/4"	Each	46	150
3'4" x 6'8" x 4 3/4"	Each	55	155
3'6" x 6'8" x 4 3/4"	Each	62	160
4'0" x 6'8" x 4 3/4"	Each	70	170
Deduct for 18 Gauge	Each	8	15
Add for 14 Gauge	Each	11	22
Add for A, B, or C Label	Each	7	12
Add for Transoms	Each	27	70
Add for Side Lights	Each	29	85
Add for Frames over 7'0" (Height)	Each	29	15
Add for Frames over 6 3/4" (Width)	Each	12	15
Add for Bolted Frame with Bolts	Each	20	30
.12 Doors (1 3/4" - 18 Gauge)			
2'6" x 6'8" x 1 3/4"	Each	47	160
2'8" x 6'8" x 1 3/4"	Each	48	170
3'0" x 6'8" x 1 3/4"	Each	50	180
3'4" x 6'8" x 1 3/4"	Each	57	190
3'6" x 6'8" x 1 3/4"	Each	64	205
4'0" x 6'8" x 1 3/4"	Each	75	220
Add for 16 Gauge	Each	10	35
Deduct for 20 Gauge	Each	7	15
Add for A Label	Each	8	15
Add for B or C Label	Each	7	15
Add for Vision Panels or Lights	Each	-	65
Add for Louvre Openings	Each	-	60
Add for Galvanizing	Each	-	40
.2 STOCK			
.21 Frames (4 3/4"-5 3/4"-6 3/4"-8 3/4"- 16 Ga)			
2'6" x 6'8" or 7'0"	Each	43	80
2'8" x 6'8" or 7'0"	Each	44	85
3'0" x 6'8" or 7'0"	Each	47	90
3'4" x 6'8" or 7'0"	Each	53	95
3'6" x 6'8" or 7'0"	Each	60	100
4'0" x 6'8" or 7'0"	Each	68	120
See above .11 for Size Change			
Add for Bolted Frame & Bolts	Each	20	30
Add for A, B, or C Label	Each	8	15
Deduct Non-Welded Frame (Knock Down)	Each	5	15
.22 Doors (1 3/8" or 1 3/4" - 18 Gauge)			
2'6" x 6'8" or 7'0"	Each	47	165
2'8" x 6'8" or 7'0"	Each	48	170
3'0" x 6'8" or 7'0"	Each	50	180
3'4" x 6'8" or 7'0"	Each	57	190
3'6" x 6'8" or 7'0"	Each	63	215
4'0" x 6'8" or 7'0"	Each	70	225
Add for A Label	Each	8	15
Add for B or C Label	Each	7	10
Add for Galvanizing	Each	-	40

Labor Above Includes Installation of Butts & Locksets

Add to All Above for:

	UNIT	LABOR	MATERIAL
Door Closures - Surface Mounted	Each	45	90
Panic Bars (Exit Device)	Each	90	430
Kick, Push and Pull Plates	Each	18	30

See 8-13 for Other Hardware

0802.0 WOOD DOORS (M) (CARPENTERS)

See 0603.3 and 0802.5 for Custom Frames

Labor includes Installation of Butts and Locksets

.1 FLUSH DOORS - Pre-Machined

No Label - Standard - Birch - 7 Ply - Paint Grade

	UNIT	LABOR	Hollow Core	Solid Core
2'0" x 6'8" or 1 3/8"	Each	45	90	100
2'4" x 6'8" or 1 3/8"	Each	46	95	105
2'6" x 6'8" or 1 3/8"	Each	47	100	110
2'8" x 6'8" or 1 3/8"	Each	48	105	115
3'0" x 6'8" or 1 3/8"	Each	50	110	120
2'0" x 6'8" or 1 3/4"	Each	46	95	110
2'4" x 6'8" or 1 3/4"	Each	47	100	115
2'6" x 6'8" or 1 3/4"	Each	48	105	120
2'8" x 6'8" or 1 3/4"	Each	49	105	125
3'0" x 6'8" or 1 3/4"	Each	52	110	130
3'4" x 6'8" or 1 3/4"	Each	56	125	145
3'6" x 6'8" or 1 3/4"	Each	60	185	210
3'8" x 6'8" or 1 3/4"	Each	65	200	220
Add for Stain Grade	Each			25
Add for 5 Ply	Each			65
Add for Jamb & Trim - Solid - Knock Down	Each			70
Add for Jamb & Trim- Veneer- Knock Down	Each			60
Add for Red Oak - Rotary	Each			20
Add for Red Oak - Plain Sliced	Each			35
Add for Architectural Grade - 7 Ply	Each			75
Deduct for Lauan	Each			15
Add for Doors over 6'8" per Inch	Each			12
Add for Vinyl Overlay	Each			45
Add for Cutouts and Special Cuts	Each			50
Add for Lite Cutouts w/Metal Frames	Each			75
Add for Private Door Eye	Each			20
Add for Wood Louvre	Each			95
Add for Transom Panels & Side Panels	Each	24		105
Add for Astragals	Each	25		40
Add for Trimming Door for Carpet	Each	20		-

Label - Birch - Paint Grade	UNIT	LABOR	MATERIAL 20-Min	MATERIAL 45-Min	MATERIAL 60-Min
2'0" x 6'8" x 1 3/4"	Each	45	120	150	155
2'4" x 6'8" x 1 3/4"	Each	46	130	160	165
2'6" x 6'8" x 1 3/4"	Each	47	135	170	175
2'8" x 6'8" x 1 3/4"	Each	48	140	200	210
3'0" x 6'8" x 1 3/4"	Each	49	145	240	250
3'4" x 6'8" x 1 3/4"	Each	55	160	260	265
3'6" x 6'8" x 1 3/4"	Each	60	225	270	275
3'8" x 6'8" x 1 3/4"	Each	65	240	285	290

See Above 0602.1 for Changes

.2 PANEL DOORS

	UNIT	LABOR	MATERIAL Pine	MATERIAL Fir	MATERIAL Oak
Exterior Entrances					
2'8" x 6'8" x 1 3/4"	Each	73	300	310	475
3'0" x 6'8" x 1 3/4"	Each	78	320	330	500
Add for Sidelights	Each	35	170	195	315
Interior					
2'6" x 6'8" x 1 3/8"	Each	55	265	245	390
2'8" x 6'8" x 1 3/8"	Each	60	275	250	400
3'0" x 6'8" x 1 3/8"	Each	65	300	260	420

0802.0 WOOD DOORS (Cont'd...)

.3 LOUVERED DOORS	UNIT	LABOR	MATERIAL Birch/Oak	Pine	Lauan
1'0" x 6'8" x 1 3/8" Pine	Each	30	350	125	-
1'3" x 6'8" x 1 3/8"	Each	33	370	130	-
1'6" x 6'8" x 1 3/8"	Each	37	390	135	-
2'0" x 6'8" x 1 3/8"	Each	45	460	160	-
2'4" x 6'8" x 1 3/8"	Each	50	460	175	-
2'6" x 6'8" x 1 3/8"	Each	50	470	215	-
2'8" x 6'8" x 1 3/8"	Each	52	480	220	-
3'0" x 6'8" x 1 3/8"	Each	62	490	240	-
Add for Half Louvered/Panel	Each				15%
Add for Prehung	Each	-	90	90	0

.4 BI-FOLD - FLUSH

	UNIT	LABOR	Birch/Oak	Pine	Lauan
2'0" x 6'8" x 1 3/8" Pine	Each	45	-	-	75
2'4" x 6'8" x 1 3/8"	Each	48	-	-	80
2'6" x 6'8" x 1 3/8"	Each	52	-	-	80
2'8" x 6'8" x 1 3/8"	Each	54	-	-	85
3'0" x 6'8" x 1 3/8"	Each	59	-	-	90
Add for Oak	Each	-	35	-	-
Add for Birch	Each	-	30	-	-
Add for Louvered Type	Each	10			80
Add for Half Louvered/Panel	Each	10			80
Add for Decorative Type	Each	10			165
Add for Prefinished Type	Each	-			20
Add for Heights over 6'8"	Each	10			4
Add for Widths over 3'0"	Each	10			5

See 0803.2 for Closet Bi-fold Doors and Prehung Stock

.5 CAFE DOORS

	UNIT	LABOR	Oak	Pine	Fir
2'6" x 3'8" x 1 1/8"	Pair	60	-	225	225
2'8" x 3'8" x 1 1/8"	Pair	65	-	235	235
3'0" x 3'8" x 1 1/8"	Pair	70	-	245	245

.6 FRENCH - with Grille Pattern

	UNIT	LABOR	Oak	Pine	Fir
2'6" x 6'8" x 1 3/8"	Each	70	430	490	300
2'8" x 6'8" x 1 3/8"	Each	72	440	500	330
3'0" x 6'8" x 1 3/8"	Each	75	450	510	450
Add for Prehung	Each	-	90	-	-

.7 DUTCH

	UNIT	LABOR	Oak	Pine	Fir
2'6" x 6'8" x 1 3/4"	Each	90	-	590	-
2'8" x 6'8" x 1 3/4"	Each	95	-	650	-
3'0" x 6'8" x 1 3/4"	Each	100	-	675	-

.8 PREHUNG DOOR UNITS - Incl. Frame, Trim, Hardware and Jambs

Exterior

	UNIT	LABOR	Birch	Pine	Fir
2'8" x 6'8" x 1 3/4"	Each	60	-	275	-
3'0" x 6'8" x 1 3/4"	Each	65	-	310	-
3'4" x 6'8" x 1 3/4"	Each	70	-	410	-
Add for Insulated	Each	10	-	-	130
Add for Threshold	Each	10	-	-	15
Add for Weatherstrip	Each	-	-	-	80
Add - Sidelites 1'0" HC	Each	20	-	-	140
Add - Sidelites 1'6" HC	Each	25	-	-	170

Interior

	UNIT	LABOR	Birch	Lauan	Oak
2'6" x 6'8" x 1 3/8"	Each	56	185	130	190
2'8" x 6'8" x 1 3/8"	Each	60	195	135	195
3'0" x 6'8" x 1 3/8"	Each	65	200	140	200

0803.0 SPECIAL DOORS

				MATERIAL			
.1	BLAST OR NUCLEAR RESISTANT						Pine
.2	CLOSET - BI-FOLDING (PREHUNG STOCK)			Oak	Birch	Lauan	& Fir
.21	Wood - Units - 1 3/8"	UNIT	LABOR	Flush	Flush	Flush	Panel
	Incl. Jamb, Csg. Track & Hdwe.						
	2 Door - 2'0" x 6'8"	Each	40	150	165	110	360
	2'6" x 6'8"	Each	44	158	175	115	380
	3'0" x 6'8"	Each	48	165	180	120	415
	4 Door - 4'0" x 6'8"	Each	53	230	240	160	600
	5'0" x 6'8"	Each	60	240	260	170	660
	6'0" x 6'8"	Each	67	265	285	190	740
	Incl. Track & Hdwe (Unfinished)						
	2 Door - 2'0" x 6'8"	Each	26	90	85	85	-
	2'6" x 6'8"	Each	38	100	90	90	-
	3'0" x 6'8"	Each	40	110	100	100	-
	4 Door - 4'0" x 6'8"	Each	43	145	130	130	-
	5'0" x 6'8"	Each	46	155	155	145	-
	6'0" x 6'8"	Each	50	175	175	155	-
	Add for Louvered Type	Each					100%

		UNIT	LABOR	MATERIAL
.22	Metal			
	Flush 2 Door - 2'0" x 6'8"	Each	33	75
	2'6" x 6'8"	Each	35	80
	3'0" x 6'8"	Each	37	85
	Flush 4 Door - 4'0" x 6'8"	Each	40	110
	5'0" x 6'8"	Each	42	115
	6'0" x 6'8"	Each	48	125
	Add for Plastic Overlay	Each	-	30
	Add for Louvered or Decorative	Each	-	10
.23	Leaded Mirror			
	4 Panel Unit - 4'0" x 6'8"	Each	75	540
	5'0" x 6'8"	Each	80	580
	6'0" x 6'8"	Each	90	635
.3	FLEXIBLE DOORS (M) (CARPENTER)			
	(See 1020 - Retractable Partitions)			
	Wood Slat (Unfinished)	SqFt	1.85	9.00
	Fabric Accordion Fold	SqFt	1.75	8.50
	Vinyl Clad	SqFt	2.00	11.50

				COST
.4	GLASS-ALL (L&M) (GLAZIERS)			
	Manual (Including Hardware)	Each	-	1,900
	Add for Center Locking	Each	-	165
	Automatic (Including Hardware & Operator)			
	Mat or Floor Unit	Each	-	5,600
	Handle Unit	Each	-	6,100
.5	HANGAR (L&M) (IRONWORKERS)			
	Biparting or Sliding - Steel			
	Example: 100' x 20' ($22 SqFt)	Each	-	50,000
	Add for Pass Doors	Each	-	1,200
	Add for Electric Operator	Each	-	1,600
	Add for Insulating	Each	-	1,700
	Tilt Up or Canopy	Each	Approx. Same	
.6	METAL COVERED FIRE & INDUSTRIAL SLIDING DOORS			
	3'0" x 7'0" - Flush	Each	-	580
	4'0" x 7'0"	Each	-	1,650
	8'0" x 8'0" - U.L. Label	Each	-	2,250
	10'0" x 10'0"	Each	-	2,800
	10'0" x 12'0"	Each	-	3,100
	12'0" x 12'0"	Each	-	3,500

0803.0 SPECIAL DOORS, Cont'd...

.7 OVERHEAD DOORS (L&M) (CARPENTERS)

	UNIT	COST 1 3/4" Wood	2" 24-Ga Steel
Commercial			
8' x 8'	Each	740.00	750.00
8' x 10'	Each	860.00	880.00
8' x 12'	Each	970.00	990.00
10' x 10'	Each	1,100.00	1,150.00
10' x 12'	Each	1,140.00	1,180.00
12' x 12'	Each	1,280.00	1,300.00
14' x 10'	Each	1,340.00	1,350.00
14' x 12'	Each	1,630.00	1,640.00
Add for Panel Door	Each		5%
Add for Low Lift	Each		75.00
Add for High Lift	Each		125.00
Add for Electric Operator	Each		500.00
Add for Insulated Type	Each		10%
Residential			
8' x 7' x 1 3/4"	Each	405.00	420.00
9' x 7' x 1 3/4"	Each	435.00	450.00
16' x 7' x 1 3/4"	Each	860.00	975.00
Add for Wood Paneled Door	Each		10%
Add for Redwood	Each		190.00
Add for Electric Operator	Each		375.00
Add for Vision Lights	Each		20.00

.8 PLASTIC LAMINATE FACED (M) (CARPENTERS)

	UNIT	LABOR	MATERIAL
1 3/8"	SqFt	3.60	10.25
1 3/4"	SqFt	3.70	10.60
Add for Edge Strip - Vertical	Each		6.70
Add for Edge Strip - Horizontal	Each		5.40
Add for Cutouts	Each		56.00
Add for Metal Frame Cutout	Each		83.00
Add for Decorative Type	SqFt		17.50
Add for Solid Colors	SqFt		10%
Add for 1-hour B Label	SqFt		4.60
Add for 1 1/2-hour B Label	SqFt		4.90
Add for Color Core	-		50%
Add for Machining for Hardware	Each		47.00

.9 REVOLVING DOOR (L&M) (GLAZIERS)

	UNIT	COST
Aluminum		
3 Leaf - 6'-6" x 6'-6"	Each	20,500.00
4 Leaf - 6'-6" x 6'-6"	Each	22,000.00
6'-0" x 7'-0"	Each	15,800.00
Automatic - 6'-6" x 10'-0"	Each	31,000.00
Stainless	Each	27,000.00
Bronze	Each	30,000.00
Painted Steel - Dark Room 32"	Each	1,650.00
36"	Each	1,750.00

0803.0 SPECIAL DOORS, Cont'd...	UNIT	LABOR	MATERIAL
.10 ROLLING DOORS & GRILLES (M) (IRONWORKERS)			
Doors - Manual - 8' x 8'	Each	370.00	1,700.00
10' x 10'	Each	440.00	1,750.00
12' x 12'	Each	430.00	2,350.00
14' x 14'	Each	680.00	3,260.00
Add for Electric Controlled	Each	-	245.00
Add for Fire Doors	Each	-	575.00
Grilles - Manual -6'8" x 3'2"	Each	160.00	685.00
6'8" x 4'2"	Each	145.00	760.00
8'0" x 3'2"	Each	180.00	800.00
8'0" x 4'2"	Each	190.00	870.00
Add for Electric Controlled	Each	-	250.00
.11 SHOWER DOORS (M) (CARPENTERS)			
28" x 66" Obscure Pattern Glass Wire	Each	50.00	135.00
Tempered Glass	Each	50.00	145.00
.12 SLIDING OR PATIO DOORS (M) (CARPENTERS)			
ONE SLIDING, ONE FIXED			
Metal (Aluminum) (Including 5/8" Glass, Threshold, Hardware and Screen)			
8'0" x 6'10"	Each	135.00	1,050.00
8'0" x 8'0"	Each	180.00	1,180.00
Wood (M) (Carpenters) Vinyl Sheathed (Including 5/8" Insulated Dbl. Temp. Glass, Weatherstripped, Hardware and Casing)			
Vinyl Clad 6' 0" x 6' 8"	Each	135.00	1,400.00
8' 0" x 6' 8"	Each	165.00	1,660.00
Pine - Prefinished 6' 0" x 6' 8"	Each	135.00	1,330.00
8' 0" x 6' 8"	Each	165.00	1,600.00
Deduct for 3/16" Safety Glass/Insulated	Each	-	165.00
Add for Tripe Glazing	Each	-	200.00
Add for Screen	Each	15.00	135.00
Add for Grilles	Each	18.00	200.00
.13 SOUND REDUCTION (M) (CARPENTERS)			
Metal	Each	150.00	1,430.00
Wood	Each	140.00	775.00
.14 TRAFFIC (M) (CARPENTERS)			
Electric Powered			
Hollow Metal - 8' x 8'	Each	350.00	4,100.00
8' x 8' High Speed	Each	940.00	9,000.00
Metal Clad (26 ga.) - 8' x 8'	Each	350.00	3,900.00
Wood	Each	300.00	3,200.00
Truck Impact - Double Acting (Including Bumpers and Vision Panels)			
Metal Clad - 8' x 8' Openings	Each	260.00	2,250.00
Rubber Plastic - 8' x 8' Openings 1/2"	Each	210.00	1,860.00
Aluminum - 8' x 8' Openings 1 3/4"	Each	230.00	2,330.00
.15 SPECIAL MADE & ENGINEERED INDUSTRIAL DOORS (L&M) (CARPENTERS)			
Four-Fold Garage Doors - Electric Operated	SqFt	-	95.00
.16 VAULT DOORS (M) (STEEL ERECTORS)			
1 hour with Frame 6' 6" x 2' 8"	Each	310.00	1,900.00
2 hour with Frame 6' 6" x 2' 8"	Each	310.00	2,500.00
6' 6" x 3' 0"	Each	365.00	2,850.00
6' 6" x 3' 4"	Each	410.00	3,100.00

0804.0 ENTRANCE DOORS, FRAMES & STORE FRONT CONSTRUCTION

Prices Include Hardware & Safety Plate Glazing

		UNIT	COST
.1	**ALUMINUM**		
.11	Stock		
	Door and Frame - 3' x 7'	Each	1,350.00
	6' x 7'	Each	2,440.00
	Window or Store Front Framing - 10'	SqFt	24.00
.12	Custom		
	Door and Frame - 3' x 7'	Each	1,660.00
	6' x 7'	Each	3,100.00
	Window or Store Front Framing - 10'	SqFt	31.00
	Add for Anodized Finish		15%
.2	**BRONZE**		
	Door and Frame - 3' x 7'	Each	3,400.00
	Window or Store Front Framing	SqFt	60.00
.3	**STAINLESS STEEL**		
	Door and Frame - 3' x 7'	Each	2,900.00
	Window or Store Front Framing	SqFt	58.00
	Add for Insulated Glass	SqFt	6.25
	Add for Exit Hardware	Each	310.00

See 0803.10 for Revolving Doors
See 0803.0 for Automatic Door Opening Assemblies

0805.0 METAL WINDOWS (M) (CARPENTERS OR STEEL ERECTORS) (SINGLE GLAZING INCLUDED)

All Example Sizes Below: 2'4" x 4'6"

		UNIT	LABOR	MATERIAL	or	UNIT	LABOR	MATERIAL
.1	**ALUMINUM WINDOWS**							
	Casement and Awning	Each	47.00	230.00		SqFt	4.40	22.00
	Sliding or Horizontal Rolling	Each	47.00	170.00		SqFt	4.40	16.00
	Double and Single-Hung or Vertical Sliding	Each	47.00	170.00		SqFt	4.45	16.00
	Projected	Each	43.00	210.00		SqFt	3.90	20.00
	Add for Screens					SqFt	.60	3.50
	Add for Storms					SqFt	1.25	5.60
	Add for Insulated Glass					SqFt	-	5.85
	Add for Bronzed Finish					SqFt	-	10%
.2	**ALUMINUM SASH**							
	Casement	Each	35.00	210.00		SqFt	3.40	19.50
	Sliding	Each	35.00	150.00		SqFt	3.40	14.00
	Single-Hung	Each	35.00	150.00		SqFt	3.40	14.00
	Projected	Each	35.00	235.00		SqFt	3.40	22.00
	Fixed	Each	35.00	120.00		SqFt	2.70	11.50
.3	**STEEL WINDOWS**							
	Double-Hung	Each	43.00	340.00		SqFt	4.00	33.00
	Projected	Each	43.00	285.00		SqFt	4.00	28.00
	Add for Insulated Glass					SqFt	-	5.00
.4	**STEEL SASH**							
	Casement	Each	36.00	240.00		SqFt	3.25	23.00
	Double-Hung	Each	36.00	320.00		SqFt	3.25	31.00
	Projected	Each	36.00	275.00		SqFt	3.25	27.00
	Fixed	Each	36.00	160.00		SqFt	3.25	16.00

0806.0 WOOD WINDOWS (M) (Carpenters)

All Units Assembled, Kiln Dry Pine, Glazed, Vinyl Clad, Weather-stripped, and No Interior Trim. <u>Top Quality</u>. Other Sizes Available. Most commonly used listed below. Deduct 10% for Medium Quality and 25% for Low Quality Woods.

		UNIT	LABOR	MATERIAL
.1	BASEMENT OR UTILITY			
	Prefinished 2'8" x 1'4"	Each	34.00	105.00
	2'8" x 1'8"	Each	41.00	112.00
	2'8" x 2'0"	Each	43.00	120.00
	Add for Double Glazing	Each	-	38.00
	Add for Storm	Each	12.00	25.00
	Add for Screen	Each	10.00	17.00
.2	CASEMENT OR AWNING - Operable Units			
	Single Clear Tempered Glass 1'8" x 3'0"	Each	45.00	250.00
	1'8" x 3'6"	Each	46.00	265.00
	2'4" x 3'6"	Each	53.00	280.00
	2'4" x 4'0"	Each	55.00	340.00
	2'4" x 5'0"	Each	60.00	375.00
	Double (2 Units) 2'10" x 3'0"	Each	65.00	440.00
	3'4" x 3'6"	Each	70.00	530.00
	4'0" x 3'0"	Each	65.00	460.00
	4'0" x 3'6"	Each	70.00	550.00
	4'0" x 5'0"	Each	76.00	660.00
	4'0" x 5'8"	Each	78.00	715.00
	Triple (3 Units, 1 Fixed) 6'0" x 3'0"	Each	75.00	650.00
	6'0" x 4'0"	Each	80.00	760.00
	6'0" x 4'6"	Each	90.00	870.00
	Add for Grille Patterns - Interior - Polycarbonate	Each	-	20.00
	Add for Grille Patterns - Exterior - Prefinished	Each	-	55.00
	Add for High Performance Glazing	Each	-	50.00
	Add for Screen - per Unit	Each	-	18.00
	Add for Extension Jambs	Each	-	24.00
	Add for Blinds	Each	-	80.00
	CASEMENT OR AWNING - Stationary & Fixed Units			
	Single 1'8" x 3'0"	Each	45.00	210.00
	1'8" x 3'6"	Each	46.00	245.00
	2'4" x 3'6"	Each	53.00	250.00
	2'4" x 4'0"	Each	55.00	310.00
	2'4" x 5'0"	Each	62.00	335.00
	2'4" x 6'0"	Each	65.00	385.00
	Picture 4'0" x 4'0"	Each	38.00	395.00
	4'0" x 4'6"	Each	42.00	430.00
	4'0" x 6'0"	Each	44.00	575.00
	6'0" x 4'0"	Each	45.00	565.00
	6'0" x 6'0"	Each	50.00	740.00
	Add for Grille Patterns - Wood - 2'4"	Each	20.00	75.00
	Add for Grille Patterns - Wood - 4'6"	Each	25.00	115.00
.3	DOUBLE HUNG (Insulating Glass with Screen)			
	2'6" x 3'0"	Each	36.00	270.00
	2'6" x 3'6"	Each	39.00	280.00
	2'6" x 4'2"	Each	42.00	300.00
	3'2" x 3'0"	Each	48.00	295.00
	3'2" x 3'6"	Each	50.00	320.00
	3'2" x 4'2"	Each	52.00	330.00
	Deduct for Primed Units	Each	4.00	25.00
	Add for Grille Patterns	Each	7.00	25.00
	Add for Triple Glazing	Each	-	60.00
	Deduct for Screen Unit	Each	16.00	65.00

0806.0 WOOD WINDOWS, Cont'd...

		UNIT	LABOR	MATERIAL
.4	GLIDER w/SCREEN & INSUL GLASS 4'0" x 3'6"	Each	50.00	750.00
	5'0" x 3'6"	Each	55.00	810.00
	4'0" x 4'0"	Each	65.00	780.00
	5'0" x 4'0"	Each	70.00	840.00
.5	PICTURE WINDOW w/CASEMENTS (2)			
	& INSULATING GLASS 9'6" x 5'0"	Each	125.00	1,570.00
	10'1" x 5'0"	Each	130.00	1,625.00
	9'6" x 5'6"	Each	125.00	1,680.00
	SINGLE GLASS AND STORM 10'3" x 4'6"	Each	110.00	1,630.00
	10'3" x 5'6"	Each	125.00	1,650.00
	10'3" x 6'6"	Each	130.00	1,725.00
	Add for High Performance Glass	Each	-	150.00
.6	CASEMENT ANGLE BAY WINDOWS, INSULATING GLASS, VINYL SHEATHED			
	30° (3 units) 5'10" x 4'2" x 2 7/8"	Each	175.00	1,250.00
	(4 units) 7'10" x 4'2" x 2 7/8"	Each	180.00	1,550.00
	(3 units) 5'10" x 5'2" x 2 7/8"	Each	180.00	1,340.00
	(4 units) 7'10" x 5'2" x 2 7/8"	Each	180.00	1,550.00
	(5 units) 9'10" x 4'2" x 2 7/8"	Each	210.00	1,800.00
	(5 units) 9'10" x 5'2" x 2 7/8"	Each	225.00	1,960.00
	45° (3 units) 5'4" x 4'2" x 2 7/8"	Each	185.00	1,280.00
	(4 units) 7'4" x 4'2" x 2 7/8"	Each	180.00	1,375.00
	(3 units) 5'4" x 5'2" x 2 7/8"	Each	185.00	1,400.00
	(4 units) 7'4" x 5'2" x 2 7/8"	Each	190.00	1,600.00
	(5 units) 9'4" x 4'2" x 2 7/8"	Each	200.00	1,850.00
	(5 units) 9'4" x 5'2" x 2 7/8"	Each	220.00	2,030.00
	Add for Screens	Each	40.00	45.00
	Add for Combination Windows	Each	-	230.00
	Deduct for Triple Glazing	Each	-	240.00
	Add for High Performance Glazing	Each	-	65.00
	Add for Bronze Glazing	Each	-	140.00
	Add for Color	Each	-	65.00
	Add for Blinds	Each	-	300.00
	Add for Bottom or Roof Skirt - Plywood	Each	85.00	350.00
	Add for Bottom or Roof Skirt - Aluminum	Each	90.00	520.00
.7	CASEMENT BOW WINDOWS, INSULATING GLASS			
	(3 units) 6'2" x 4'2"	Each	260.00	1,110.00
	(4 units) 8'2" x 4'2"	Each	320.00	1,420.00
	(3 units) 6'2" x 5'2"	Each	270.00	1,340.00
	(4 units) 8'2" x 5'2"	Each	340.00	1,620.00
.8	90° CASEMENT BOX BAY WINDOWS, INSULATING GLASS			
	4'8" x 4'2"	Each	340.00	1,530.00
	6'8" x 4'2"	Each	415.00	1,950.00
	4'8" x 5'2"	Each	350.00	1,560.00
	6'8" x 5'2"	Each	445.00	2,000.00
	Add for Screens - per Unit	Each	40.00	65.00
	Deduct for Double Glazing - per Unit	Each	-	35.00
	Add for High Performance Glazing	Each	-	70.00

		UNIT	LABOR	STATIONARY	VENT
.9	ROOF WINDOWS - INSULATING GLASS	Each			
	1'10" x 3'10"	Each	160.00	360.00	575.00
	2'4" x 3'10"	Each	170.00	430.00	650.00
	3'8" x 3'10"	Each	185.00	520.00	770.00
	Add for Shingle Flashing	Each	35.00	75.00	70.00
	Add for High Performance Glass	Each	-	80.00	75.00

		UNIT	LABOR	MATERIAL
.10	CIRCLE TOPS - 2'4"	Each	95.00	480.00
	4'0"	Each	105.00	625.00
	4'8"	Each	110.00	730.00
	6'0"	Each	120.00	1,140.00

		UNIT	LABOR	MATERIAL
0807.0	**SPECIAL WINDOWS (M) (Carpenters/ Steel Workers)**			
.1	LIGHT-PROOF WINDOWS	SqFt	4.80	27.00
.2	PASS WINDOWS	SqFt	4.10	20.00
.3	DETENTION WINDOWS			
	Aluminum	SqFt	4.25	27.00
	Steel	SqFt	4.15	26.00
.4	VENETIAN BLIND WINDOWS (ALUMINUM)	SqFt	3.85	23.00
.5	SOUND-CONTROL WINDOWS	Each	3.65	17.00
0808.0	**DOOR AND WINDOW ACCESSORIES**			
.1	STORMS AND SCREENS (CARPENTERS)			
	Windows			
	Screen Only - Wood - 3' x 5'	Each	14.50	66.00
	Aluminum - 3' x 5'	Each	15.50	80.00
	Storm & Screen Combination - Aluminum	Each	17.50	95.00
	Doors			
	Screen Only - Wood - 3' x 6' - 8'	Each	26.00	165.00
	Aluminum	Each	24.00	185.00
	Storm & Screen Combination - Aluminum	Each	28.00	260.00
	Wood 1 1/8"	Each	30.00	240.00
.2	DETENTION SCREENS (M) (CARPENTERS)			
	Example: 4' 0" x 7' 0"	Each	75.00	470.00
.3	DOOR OPENING ASSEMBLIES (L&M) (GLAZIERS)			
	Floor or Overhead Electric Eye Units			
	Swing - Single 3' x 7' Door - Hydraulic	Each	-	3,550.00
	Double 6' x 7' Door - Hydraulic	Each	-	5,000.00
	Sliding - Single 3' x 7' Door - Hydraulic	Each	-	4,600.00
	Double 5' x 7' Door - Hydraulic	Each	-	6,200.00
	Industrial Door - 10' x 8	Each	-	6,000.00
.4	SHUTTERS - FOLDING (CARPENTERS)			
	16" x 1 1/8" x 48"	Pair	15.00	80.00
	x 60"	Pair	17.00	90.00
	x 72"	Pair	20.00	100.00

		UNIT	COST
0809.0	**CURTAIN WALL SYSTEMS (L&M) (IRONWORKERS)**		
.1	STRUCTURAL OR VERTICAL TYPE		
	Aluminum Tube Frame, Panel, Sills		
	and Mullions - No glazing		
	Example: 4' Panel, 7' Fixed	SqFt	40.00 to 120.00
.2	PANEL WALL OR HORIZONTAL TYPE		
	Aluminum Tube Frame, Panel, Sills		
	and Mullions - No glazing		
	Example: 1' Panel, 2' Oper., 4' Fixed	SqFt	35.00 to 95.00
.3	ARCHITECTURAL PANELS, INSULATED		
	Included in Above Prices		
	Metal		
	Baked Enamel on Steel - 24 ga	SqFt	12.50
	on Aluminum - 24 ga	SqFt	13.00
	Porcelain Enamel on Steel - 24 ga	SqFt	14.10
	on Aluminum - 24 ga	SqFt	14.50
	Exposed Aggregate - Resin	SqFt	11.00
	Fiberglass	SqFt	14.75
	Spandrel Glass	SqFt	13.00

See 0812 for Glazing Costs to be Added to Above

0810.0 FINISH HARDWARE (M) (CARPENTERS)

		EACH	LABOR	MATERIAL Painted	MATERIAL Bronze	MATERIAL Chrome
.1	BUTTS					
	3" x 3"	Pair	14.00	7.70	8.70	11.25
	3 1/2" x 3 1/2"	Pair	14.50	8.90	8.90	12.30
	4" x 4"	Pair	15.50	11.30	11.30	12.80
	4 1/2" x 4 1/2"	Pair	16.00	15.50	15.50	20.90
	4" x 4" Ball Bearing	Pair	17.00	29.50	29.80	38.00
	4 1/2" x 4 1/2" Ball Bearing	Pair	18.00	33.30	34.00	40.00

		UNIT	LABOR	MATERIAL
.2	CATCHES - ROLLER	Each	13.00	13.30
.3	CLOSURES - SURFACE MOUNTED - 3' - 0" Door	Each	42.00	87.00
	3' - 4"	Each	43.00	92.00
	3' - 8"	Each	44.00	95.00
	4' - 0"	Each	49.00	103.00
	CLOSURES - CONCEALED OVERHEAD - Interior	Each	70.00	144.00
	Exterior	Each	75.00	235.00
	Add for Fusible Link - Electric	Each	18.00	148.00
	CLOSURES - FLOOR HINGES - Interior	Each	100.00	460.00
	Exterior	Each	100.00	480.00
	Add for Hold Open Feature	Each	-	72.00
	Add for Double Acting Feature	Each	-	235.00
.4	DEAD BOLT LOCK - Cylinder - Outside Key	Each	32.00	112.00
	Cylinder - Double Key	Each	37.00	122.00
	Flush - Push/ Pull	Each	20.00	20.00
.5	EXIT DEVICES (PANIC) - Surface	Each	90.00	460.00
	Mortise Lock	Each	95.00	520.00
	Concealed	Each	130.00	1,650.00
	(Handrop) Automatic	Each	105.00	1,150.00
.6	HINGES, SPRING (PAINTED) - 6" Single Acting	Each	25.00	57.00
	6" Double Acting	Each	25.00	73.00
.7	LATCHSETS - Bronze or Chrome	Each	33.00	115.00
	Stainless Steel	Each	33.00	133.00
.8	LOCKSETS - Mortise - Bronze or Chrome - H.D.	Each	30.00	170.00
	S.D.	Each	28.00	123.00
	Stainless Steel	Each	28.00	230.00
	Cylindrical - Bronze or Chrome	Each	32.00	138.00
	Stainless Steel	Each	32.00	175.00
.9	LEVER HANDICAP - Latch Set	Each	30.00	143.00
	Lock Set	Each	30.00	180.00
.10	PLATES - Kick - 8" x 34" - Aluminum	Each	18.00	18.00
	Bronze	Each	19.00	38.00
	Push - 6" x 15"- Aluminum	Each	17.00	18.00
	Bronze	Each	18.00	31.00
	Push & Pull Combination - Aluminum	Each	20.00	43.00
	Bronze	Each	20.00	72.00
.11	STOPS AND HOLDERS			
	Holder - Magnetic (No Electric)	Each	40.00	105.00
	Bumper	Each	23.00	17.00
	Overhead - Bronze, Chrome or Aluminum	Each	23.00	58.00
	Wall Stops	Each	18.00	13.00
	Floor Stops	Each	18.00	14.00
	See 0808.3 for Automatic Openers & Operators			

0811.0 WEATHERSTRIPPING (L&M) (CARPENTERS)

		UNIT	LABOR	MATERIAL
.1	ASTRAGALS - Aluminum - 1/8" x 2"	Each	18.00	12.00
	Painted Steel	Each	18.00	12.00
.2	DOORS (WOOD) - Interlocking	Each	30.00	28.00
	Spring Bronze	Each	25.00	45.00
	Add for Metal Doors	Each	25.00	7.50
.3	SWEEPS - 36" WOOD DOORS - Aluminum	Each	13.00	12.50
	Vinyl	Each	13.00	8.70
.4	THRESHOLDS - Aluminum - 4" x 1/2"	Each	17.00	14.50
	Bronze - 4" x 1/2"	Each	18.00	27.50
	5 1/2" x 1/2"	Each	19.00	32.00
.5	WINDOWS (WOOD) - Interlocking	Each	33.00	23.00
	Spring Bronze	Each	28.00	40.00

0812.0	GLASS AND GLAZING (L&M) (Glaziers)	UNIT	COST
.1	BEVELED GLASS - 1/4" x 1/2" Bevel	SqFt	108.00
.2	COATED - Used with Insulated Glass and Heat		
	Absorber and Light Reflector	SqFt	25.00
.3	HEAT ABSORBING AND SOLAR REJECTING COMBINATIONS of Tinted, Coated and Clear Glass in Insulated Glass Form with 1/4" or 1/2" Air Space. Usual thickness with 1/4" air space, 9/16", 11/16" and 1 3/16". With 1/2" air space, 13/16", 15/16" and 1 /16". Factory priced. Greatly variable.	SqFt	20.00
.4	INSULATED GLASS		
	1/2" and 5/8" Welded	SqFt	14.00
	1" Clear	SqFt	15.50
	Tempered Insulated	SqFt	25.00
	Triple Insulated	SqFt	21.50
	Add for Tinting	SqFt	10%
.5	LAMINATED (SAFETY AND SOUND)		
	7/32"	SqFt	13.00
	1/4"	SqFt	15.50
	3/8"	SqFt	16.50
	1/2"	SqFt	24.00
	1 3/16", 1 1/2", 1 3/4" and 2" Bulletproof	SqFt	50.00
.6	MIRROR GLASS		
	Copper Plated Back and Polished Edges	SqFt	12.00
	Unfinished	SqFt	7.50
.7	PATTERNED OR ROUGH (OBSCURE)		
	1/8"	SqFt	6.00
	7/32"	SqFt	6.25
.8	PLASTIC GLASS (SAFETY)		
	Acrylic - 1/8" .125"	SqFt	10.25
	3/16" .187"	SqFt	12.50
	1/4" .250"	SqFt	13.00
	5/16" .312"	SqFt	15.25
	3/8" .375"	SqFt	19.00
	Polycarbonates - Add to Above	SqFt	50%
.9	PLATE - CLEAR		
	1/4" (See .4 for Tinted)	SqFt	6.75
	3/8"	SqFt	9.25
	1/2"	SqFt	14.50
	5/8"	SqFt	17.00
	3/4"	SqFt	19.60
.10	SANDBLASTED - 3/16"	SqFt	8.50
.11	SHEET GLASS		
	Light Sheet or Window Glass - Double Strength	SqFt	4.20
	Heavy Sheet or Crystal	SqFt	5.60
.12	STAINED GLASS - Abstract	SqFt	75.00
	Symbolism	SqFt	110.00
	Figures	SqFt	250.00
.13	STRUCTURAL GLASS - 1/4"	SqFt	11.00
.14	TEMPERED (SAFETY PLATE)		
	1/4"	SqFt	7.75
	3/8"	SqFt	10.00
	1/2"	SqFt	21.50
	5/8"	SqFt	25.00
.15	TINTED PLATE - LIGHT REFLECTIVE		
	1/4"	SqFt	8.25
	3/8"	SqFt	11.25
	1/2"	SqFt	19.50
.16	WIRED (SAFETY)		
	1/4" - Clear	SqFt	12.50
	Obscure	SqFt	9.00

The costs below are average, priced as total contractor / subcontractor costs with an overhead and fee of 10% included. Units include fasteners, tools and equipment, fringe benefits, 30% taxes and insurance on labor, and 5% General Conditions.

0801.0 HOLLOW METAL (Installation of Butts and Locksets included but no hardware material)

	UNIT	COST
.1 CUSTOM FRAMES (16 ga)		
2'6" x 6'8" x 4 3/4"	Each	215
2'8" x 6'8" x 4 3/4"	Each	230
3'0" x 6'8" x 4 3/4"	Each	245
3'4" x 6'8" x 4 3/4"	Each	265
Add for A, B or C Label	Each	30
Add for Side Lights	Each	120
Add for Frames over 7'0" - Height	Each	50
Add for Frames over 6 3/4" - Width	Each	45
CUSTOM DOORS (1 3/8" or 1 3/4") (18 ga)		
2'6" x 6'8" x 1 3/4"	Each	250
2'8" x 6'8" x 1 3/4"	Each	265
3'0" x 6'8" x 1 3/4"	Each	280
3'4" x 6'8" x 1 3/4"	Each	280
Add for 16 ga	Each	55
Add for B and C Label	Each	30
Add for Vision Panels or Lights	Each	65
.2 STOCK FRAMES (16 ga)		
2'6" x 6'8" or 7'0" x 4 3/4"	Each	150
2'8" x 6'8" or 7'0" x 4 3/4"	Each	155
3'0" x 6'8" or 7'0" x 4 3/4"	Each	165
3'4" x 6'8" or 7'0" x 4 3/4"	Each	180
Add for A, B or C Label	Each	30
STOCK DOORS (1 3/8" or 1 3/4" - 18 ga)		
2'6" x 6'8" or 7'0"	Each	240
2'8" x 6'8" or 7'0"	Each	255
3'0" x 6'8" or 7'0"	Each	270
3'4" x 6'8" or 7'0"	Each	295
Add for B or C Label	Each	25

0802.0 WOOD DOORS (Installation of Butts and Locksets included but no hardware material)

	UNIT	Hollow Core	Solid Core
.1 FLUSH DOORS - No Label - Paint Grade			
Birch - 7 Ply 2'4" x 6'8" x 1 3/8"	Each	160	175
2'6" x 6'8" x 1 3/8"	Each	165	180
2'8" x 6'8" x 1 3/8"	Each	170	190
3'0" x 6'8" x 1 3/8"	Each	175	200
2'4" x 6'8" x 1 3/4"	Each	170	190
2'6" x 6'8" x 1 3/4"	Each	175	195
2'8" x 6'8" x 1 3/4"	Each	180	210
3'0" x 6'8" x 1 3/4"	Each	190	215
3'4" x 6'8" x 1 3/4"	Each	195	235
3'6" x 6'8" x 1 3/4"	Each	205	320
Add for Stain Grade	Each		30
Add for 5 Ply	Each		65
Add for Jamb & Trim - Solid - Knock Down	Each		70
Add for Jamb & Trim - Veneer - Knock Down	Each		45
Add for Red Oak - Rotary Cut	Each		25
Add for Red Oak - Plain Sliced	Each		35
Deduct for Lauan	Each		20
Add for Architectural Grade - 7 Ply	Each		70
Add for Vinyl Overlay	Each		50
Add for 7'-0" Doors	Each		20
Add for Lite Cutouts w/Metal Frame	Each		80
Add for Wood Louvres	Each		110
Add for Transom Panels & Side Panels	Each		135

0802.0 WOOD DOORS, Cont'd...

.1 FLUSH DOORS, Contd...

Label - 1 3/4" - Paint Grade

	UNIT	20-Min	45-Min	60-Min
2'6" x 6'8" - Birch	Each	210	245	250
2'8" x 6'8"	Each	220	255	260
3'0" x 6'8"	Each	235	270	275
3'4" x 6'8"	Each	255	290	295
3'6" x 6'8"	Each	305	305	315

.2 PANEL DOORS

	UNIT	Pine	Fir	Birch/ Oak	Lauan
Exterior - 1 3/4"					
2'8" x 6'8"	Each	435	445	630	-
3'0" x 6'8"	Each	465	475	660	-
Interior - 1 3/8"					
2'6" x 6'8"	Each	375	350	500	-
2'8" x 6'8"	Each	385	360	515	-
3'0" x 6'8"	Each	415	370	540	-
Add for Sidelights	Each	245	370	400	

.3 LOUVERED DOORS - 1 3/8"

	UNIT	Pine	Fir	Birch/ Oak	Lauan
2'0" x 6'8"	Each	250	250	540	-
2'6" x 6'8"	Each	300	300	565	-
2'8" x 6'8"	Each	310	310	575	-
3'0" x 6'8"	Each	320	320	595	-

.4 BI-FOLD - 2 DOOR w/ HDWE - 1 3/8"

	UNIT	Pine	Fir	Birch/ Oak	Lauan
2'0" x 6'8" - Flush	Each	-	-	190	165
2'4" x 6'8"	Each	-	-	200	175
2'6" x 6'8"	Each	-	-	210	185
Add for Prefinished	Each				20

See 0804.22 for 1 1/8" Closet Bi-folds

.5 CAFE DOORS - 1 1/8" - Pair

	UNIT	Pine	Fir	Birch/ Oak	Lauan
2'6" x 3'8"	Each	355	-	-	-
2'8" x 3'8"	Each	370	-	-	-
3'0" x 3'8"	Each	385	-	-	-

.6 FRENCH DOORS - 1 3/8"

	UNIT	Pine	Fir	Birch/ Oak	Lauan
2'6" x 3'8"	Each	595	465	645	-
2'8" x 3'8"	Each	605	475	655	-
3'0" x 3'8"	Each	625	485	665	-

.7 DUTCH DOORS

	UNIT	Pine	Fir	Birch/ Oak	Lauan
2'6" x 3'8"	Each	755	-	-	-
2'8" x 3'8"	Each	815	-	-	-
3'0" x 3'8"	Each	865	-	-	-

.8 PREHUNG DOOR UNITS (INCL. FRAME, TRIM, HARDWARE, and SPLIT JAMBS)

	UNIT	Pine	Fir	Birch/ Oak	Lauan
Exterior Entrance - 1 3/4"					
Panel - 2'8" x 6'8"	Each	415	-	-	-
3'0" x 6'8"	Each	445	-	-	-
3'0" x 7'0"	Each	485	-	-	-
Add for Insulation	Each	175			
Add for Side Lights	Each	185			
Interior - 1 3/8"					
Flush, H.C. - 2'6" x 6'8"	Each	-	-	305	220
2'8" x 6'8"	Each	-	-	315	225
3'0" x 6'8"	Each	-	-	330	230
Add for Int. Panel Door	Each				200

0803.0   SPECIAL DOORS

.2   CLOSET BI-FOLDING - PREHUNG
STOCK AND UNFINISHED

	UNIT	Oak Flush	Birch Flush	Lauan Flush	Pine Panel
Wood - 2 Door - 2'0" x 6'8" x 1 3/8"	Each	225	260	195	435
2'6" x 6'8" x 1 3/8"	Each	245	280	200	460
3'0" x 6'8" x 1 3/8"	Each	258	330	210	510
4 Door - 4'0" x 6'8" x 1 3/8"	Each	300	340	265	735
5'0" x 6'8" x 1 3/8"	Each	315	355	300	800
6'0" x 6'8" x 1 3/8"	Each	350	370	315	870

	UNIT	COST
Add for Prefinished -	Each	$30.00
Add for Jambs and Casings -	Each	$65.00
Metal - 2 Door - 2'0" x 6'8"	Each	135
2'6" x 6'8"	Each	140
3'0" x 6'8"	Each	160
4 Door - 4'0" x 6'8"	Each	185
5'0" x 6'8"	Each	195
6'0" x 6'8"	Each	210
Add for Plastic Overlay	Each	35
Add for Louvre or Decorative Type	Each	15
Leaded Mirror (Based on 2-Panel Unit)		
4'0" x 6'8"	Each	675
5'0" x 6'8"	Each	725
6'0" x 6'8"	Each	790

.3   FLEXIBLE DOORS (M) (CARPENTERS)

	UNIT	COST
Wood Slat (Unfinished)	SqFt	1,360
Fabric Accordion Fold	SqFt	1,300
Vinyl Clad	SqFt	1,700

.8   PLASTIC LAMINATE FACED (M) (CARPENTERS) - 1 3/8"

	UNIT	COST
PLASTIC LAMINATE FACED (M) (CARPENTERS) - 1 3/8"	SqFt	1,750
1 3/4"	SqFt	1,780
Add for Decorative Type	SqFt	1,850
Add for Label Doors	SqFt	4.50
Add for Edge Strip - Vertical and Horizontal	Each	8.50
Add for Machining for Hardware	Each	4,500

.10   ROLLING - DOORS & GRILLES

	UNIT	COST
Doors - 8' x 8'	Each	2,300
10' x 10'	Each	2,500
Grilles - 6'-8" x 3'-2"	Each	875
6'-8" x 4'-2"	Each	980

.11   SHOWER DOORS - 28" x 66" (CARPENTERS)   Each   225

.12   SLIDING OR PATIO DOORS (M) (CARPENTERS)
Metal (Aluminum) - Including Glass Thresholds & Screen

	UNIT	COST
Opening 8'0" x 6'8"	Each	1,475
8'0" x 8'0"	Each	1,580
Wood		
Vinyl Clad 6'0" x 6'10"	Each	1,730
8'0" x 6'10"	Each	1,950
Pine - Prefinished 6'0" x 6'10"	Each	1,780
8'0" x 6'10"	Each	1,960
Add for Grilles	Each	280
Add for Triple Glazing	Each	200
Add for Screen	Each	135

.13   SOUND REDUCTION - Metal

	UNIT	COST
SOUND REDUCTION - Metal	Each	1,700
Wood	Each	1,050

.14   TRAFFIC DOORS (CARPENTERS)

	UNIT	COST
Electric Powered - Hollow Metal - 8' x 8' Opening	Each	4,900
Wood - 8' x 8' Opening	Each	4,800
Truck Impact - Metal Clad - 8' x 8' Opening	Each	2,900
(Double Acting) - Rubber - 8' x 8' Opening	Each	2,350
Add for Frames	Each	740

.16   VAULT DOORS - 6'6" x 2'2" with Frame - 2 hour   Each   2,950

0806.0 WOOD WINDOWS (M) (Carpenters) - All Units Assembled, Kiln Dry Pine, Glazed, Vinyl Clad, Weatherstripped, and Top Quality. Many Other Sizes Available.

		UNIT	COST
.1	BASEMENT OR UTILITY		
	Prefinished with Screen 2'8" x 1'4"	Each	175
	2'8" x 2'0"	Each	200
	Add for Double Glazing or Storm	Each	45
.2	CASEMENT OR AWNING		
	Operating Units - Insulating Glass		
	Single 2'4" x 4'0"	Each	400
	2'4" x 5'0"	Each	430
	Double w/Screen 4'0" x 4'0" (2 units)	Each	610
	4'0" x 5'0" (2 units)	Each	810
	Triple w/Screen 6'0" x 4'0" (3 units)	Each	960
	6'0" x 5'0" (3 units)	Each	1,110
	Fixed Units - Insulating Glass		
	Single 2'4" x 4'0"	Each	370
	2'4" x 5'0"	Each	435
	Picture 4'0" x 4'6"	Each	505
	4'0" x 6'0"	Each	660
.3	DOUBLE HUNG - & Insul. Glass w/Screen - 2'6" x 3'6"	Each	365
	2'6" x 4'2"	Each	380
	3'2" x 3'6"	Each	390
	3'2" x 4'2"	Each	450
	Add for Triple Glazing	Each	60
.4	GLIDER - & Insul. Glass w/Screen - 4'0" x 3'6"	Each	890
	5'0" x 4'0"	Each	1,050
.5	PICTURE WINDOWS w/CASEMENTS - & Insul. Glass w/Screen		
	9'6" x 4'10"	Each	1,880
	9'6" x 5'6"	Each	2,000
.6	CASEMENT ANGLE BAY WINDOWS-Insulating Glass w/Screens		
	30° - 5'10" x 4'2" - (3 units)	Each	1,450
	7'10" x 4'2" - (4 units)	Each	1,850
	5'10" x 5'2" - (3 units)	Each	1,600
	7'10" x 5'2" - (4 units)	Each	1,800
	45° - 5'4" x 4'2" - (3 units)	Each	1,500
	7'4" x 4'2" - (4 units)	Each	1,630
	5'4" x 5'2" - (3 units)	Each	1,750
	7'4" x 5'2" - (4 units)	Each	1,950
.7	CASEMENT BOW WINDOWS - Insulating Glass		
	6'2" x 4'2" - (3 units)	Each	1,500
	8'2" x 4'2" - (4 units)	Each	2,000
	6'2" x 5'2" - (3 units)	Each	1,650
	8'2" x 5'2" - (4 units)	Each	2,150
	Add for Triple Glazing - per unit	Each	60
	Add for Bronze Glazing - per unit	Each	60
	Deduct for Primed Only - per unit	Each	15
	Add for Screens - per unit	Each	18
.8	90° CASEMENT BOX BAY WINDOWS - Insulating Glass 4'8" x 4'2"	Each	2,000
	6'8" x 4'2"	Each	2,675
	6'8" x 5'2"	Each	2,800

		UNIT	STATIONARY	MOVABLE
.9	SKY OR ROOF WINDOWS 1'10" x 3'10"	Each	560	810
	2'4" x 3'10"	Each	660	950
	3'8" x 3'10"	Each	790	1,140
.10	CIRCULAR TOPS & ROUNDS - 4'0"	Each	-	630
	6'0"	Each	-	1,340

0805.0 METAL WINDOWS (M) (CARPENTERS OR STEEL ERECTORS) (SINGLE GLAZING INCLUDED)

All Example Sizes Below: 2'4" x 4'6"

		UNIT	COST	UNIT	COST
.1	ALUMINUM WINDOWS				
	Casement and Awning	Each	315	SqFt	30
	Sliding or Horizontal	Each	240	SqFt	23
	Double and Single-Hung or Vertical Sliding	Each	270	SqFt	26
	Projected	Each	315	SqFt	30
	Add for Screens	Each	47	SqFt	4.50
	Add for Storms	Each	84	SqFt	8.00
	Add for Insulated Glass	Each	68	SqFt	6.50
.2	ALUMINUM SASH				
	Casement	Each	295	SqFt	28
	Sliding	Each	242	SqFt	23
	Single-Hung	Each	242	SqFt	23
	Projected	Each	200	SqFt	19
	Fixed	Each	210	SqFt	20
.3	STEEL WINDOWS				
	Double-Hung	Each	440	SqFt	42
	Projected	Each	420	SqFt	40
.4	STEEL SASH				
	Casement	Each	326	SqFt	31
	Double-Hung	Each	420	SqFt	40
	Projected	Each	380	SqFt	36
	Fixed	Each	288	SqFt	27

0807.0 SPECIAL WINDOWS (M) (CARPENTERS/ STEEL WORKERS)

		UNIT	COST
.1	LIGHT-PROOF WINDOWS	SqFt	36
.2	PASS WINDOWS	SqFt	26
.3	DETENTION WINDOWS	SqFt	38
.4	VENETIAN BLIND WINDOWS (ALUMINUM)	SqFt	31
.5	SOUND-CONTROL WINDOWS	SqFt	30

0808.0 DOOR AND WINDOW ACCESSORIES

		UNIT	COST
.1	STORMS AND SCREENS (CARPENTERS)		
	Windows		
	Screen Only - Wood - 3' x 5'	Each	95
	Aluminum - 3' x 5'	Each	115
	Storm & Screen Combination - Aluminum	Each	130
	Doors		
	Screen Only - Wood	Each	225
	Aluminum - 3' x 6' - 8'	Each	235
	Storm & Screen Combination - Aluminum	Each	285
	Wood 1 1/8"	Each	315
.2	DETENTION SCREENS (M) (CARPENTERS)		
	Example: 4' 0" x 7' 0"	Each	655
.3	DOOR OPENING ASSEMBLIES (L&M) (GLAZIERS)		
	Floor or Overhead Electric Eye Units		
	Swing - Single 3' x 7' Door - Hydraulic	Each	3,800
	Double 6' x 7' Door - Hydraulic	Each	5,700
	Sliding - Single 3' x 7' Door - Hydraulic	Each	5,000
	Double 5' x 7' Doors- Hydraulic	Each	6,700
	Industrial Doors - 10' x 8'	Each	6,800
.4	SHUTTERS (CARPENTERS)		
	16" x 1 1/8" x 48"	Pair	110.00
	16" x 1 1/8" x 60"	Pair	120.00
	16" x 1 1/8" x 72"	Pair	125.00

0810.0 FINISH HARDWARE (M) (CARPENTERS)

			COST		
		UNIT	Painted	Bronze	Chrome
.1	BUTTS				
	3" x 3"	Pair	30	32	35
	3 1/2" x 3 1/2"	Pair	32	33	37
	4" x 4"	Pair	34	35	39
	4 1/2" x 4 1/2"	Pair	37	40	47
	4" x 4" Ball Bearing	Pair	57	59	66
	4 1/2" x 4 1/2" Ball Bearing	Pair	62	67	70

		UNIT	COST
.2	CATCHES - ROLLER	Each	35
.3	CLOSURES - SURFACE MOUNTED - 3' - 0" Door	Each	170
	3' - 4"	Each	177
	3' - 8"	Each	180
	4' - 0"	Each	192
	CLOSURES - CONCEALED OVERHEAD - Interior	Each	275
	Exterior	Each	390
	Add for Fusible Link - Electric	Each	195
	CLOSURES - FLOOR HINGES - Interior	Each	660
	Exterior	Each	685
	Add for Hold Open Feature	Each	80
	Add for Double Acting Feature	Each	260
.4	DEAD BOLT LOCK - Cylinder - Outside Key	Each	175
	Cylinder - Double Key	Each	200
	Flush - Push/ Pull	Each	55
.5	EXIT DEVICES (PANIC) - Surface	Each	660
	Mortise Lock	Each	780
	Concealed	Each	1,880
	Handicap (ADA) Automatic	Each	1,375
.6	HINGES, SPRING (PAINTED) - 6" Single Acting	Each	100
	6" Double Acting	Each	125
.7	LATCHSETS - Bronze or Chrome	Each	175
	Stainless Steel	Each	210
.8	LOCKSETS - Mortise - Bronze or Chrome - H.D.	Each	225
	Mortise - Bronze or Chrome - S.D.	Each	190
	Stainless Steel	Each	315
	Cylindrical - Bronze or Chrome	Each	205
	Stainless Steel	Each	255
.9	LEVER HANDICAP - Latch Set	Each	205
	Lock Set	Each	260
.10	PLATES - Kick - 8" x 34" - Aluminum	Each	80
	Bronze	Each	70
	Push - 6" x 15" - Aluminum	Each	45
	Bronze	Each	65
	Push & Pull Combination - Aluminum	Each	80
	Bronze	Each	115
.11	STOPS AND HOLDERS		
	Holder - Magnetic (No Electric)	Each	185
	Bumper	Each	45
	Overhead - Bronze, Chrome or Aluminum	Each	95
	Wall Stops	Each	30
	Floor Stops	Each	40
	See 0808.3 for Automatic Openers & Operators		

0811.0 WEATHERSTRIPPING (L&M) (CARPENTERS)

		UNIT	COST
.1	ASTRAGALS - Aluminum - 1/8" x 2"	Each	40
	Painted Steel	Each	40
.2	DOORS (WOOD) - Interlocking	Each	75
	Spring Bronze	Each	85
	Add for Metal Doors	Each	40
.3	SWEEPS - 36" WOOD DOORS - Aluminum	Each	35
	Vinyl	Each	25
.4	THRESHOLDS - Aluminum - 4" x 1/2"	Each	40
	Bronze 4" x 1/2"	Each	60
	5 1/2" x 1/2"	Each	65
.5	WINDOWS (WOOD) - Interlocking	Each	67
	Spring Bronze	Each	85

		PAGE
0901.0	**LATH AND PLASTER (CSI 09100)**	**9-3**
.1	LATHING and FURRING	9-3
.11	Metal Lath	9-3
.12	Gypsum Lath	9-3
.13	Metal Studs	9-3
.14	Metal Trim	9-3
.2	PLASTERING	9-3
.21	Gypsum	9-3
.22	Acoustical	9-3
.23	Vermiculite	9-3
.24	Stucco or Cement Plaster	9-3
.25	Fireproofing (Sprayed On)	9-3
.26	Thin Set or Veneer	9-3
.27	Exterior Wall System	9-3
0902.0	**GYPSUM DRYWALL (CSI 09250)**	**9-4**
.1	BACKER BOARD	9-4
.2	FINISH BOARD	9-4
.3	VINYL COATED BOARD	9-4
.4	LINER PANELS	9-4
.5	METAL STUD PARTITIONS	9-4
.6	FURRING CHANNELS	9-4
.7	METAL TRIM	9-4
.8	DRYWALL TAPING and SANDING	9-5
.9	TEXTURING	9-5
0903.0	**TILEWORK (CSI 09300)**	**9-6**
.1	CERAMIC TILE	9-6
.2	QUARRY TILE	9-6
0904.0	**TERRAZZO (CSI 09400)**	**9-6**
.1	CEMENT TERRAZZO	9-6
.2	CONDUCTIVE TERRAZZO	9-6
.3	EPOXY TERRAZZO	9-6
.4	RUSTIC TERRAZZO	9-6
0905.0	**ACOUSTICAL TREATMENT (CSI 09500)**	**9-7**
.1	MINERAL TILE	9-7
.2	PLASTIC COVERED TILE	9-7
.3	METAL PAN TILE	9-7
.4	CLASS A TILE	9-7
.5	FIRE RATED DESIGN MINERAL TILE	9-7
.6	GLASS FIBRE BOARD TILE	9-7
.7	SOUND CONTROL PANELS and BAFFLES	9-7
.8	ACOUSTICAL WALL PANELS	9-7
.9	LINEAR METAL CEILINGS	9-7
0906.0	**VENEER STONE AND BRICK (Floors, Walls, Stools, Stairs)**	**9-8**
.1	LIMESTONE	9-8
.11	CUT STONE	9-8
.12	FLAGSTONE	9-8
.2	MARBLE	9-8
.3	SLATE	9-8
.4	SANDSTONE	9-8
.5	GRANITE	9-8
.6	SYNTHETIC MARBLE	9-8

		PAGE
0907.0	**SPECIAL FLOORING (CSI 09700)**	**9-8**
.1	MAGNESIUM OXYCHLORIDE (Seamless)	9-8
.2	ELASTOMERIC	9-8
.3	SYNTHETIC	9-8
.4	ASPHALTIC MASTIC	9-8
.5	BRICK FLOORING	9-8
.6	PRECAST PANELS	9-8
.7	NEOPRENE	9-8
0908.0	**RESILIENT FLOORING (CSI 09650)**	**9-9**
.1	ASPHALT	9-9
.2	CORK	9-9
.3	RUBBER	9-9
.4	VINYL - COMPOSITION TILE	9-9
.5	VINYL - TILE	9-9
.6	VINYL - SHEET	9-9
0909.0	**CARPETING (CSI 09680)**	**9-9**
.1	NYLON	9-9
.2	WOOL	9-9
.3	OLEFIN	9-9
.4	CARPET TILE	9-9
.5	INDOOR-OUTDOOR	9-9
0910.0	**WOOD FLOORING (CSI 09550)**	**9-10**
.1	STRIP and PLANK FLOORING	9-10
.2	WOOD PARQUET FLOORING	9-10
.3	RADIANT TREATED FLOORING	9-10
.4	RESILIENT WOOD FLOOR SYSTEMS	9-10
.5	WOOD BLOCK INDUSTRIAL FLOORS	9-10
.6	FINISHING (SANDING, STAINING, SEALING & WAXING)	9-10
0911.0	**SPECIAL COATINGS (CSI 09800)**	**9-11**
.1	CEMENTITOUS (GLAZED CEMENT)	9-11
.2	EPOXIES	9-11
.3	POLYESTERS	9-11
.4	VINYL CHLORIDES	9-11
.5	AGGREGATE to EPOXY COATING	9-11
.6	SPRAYED FIREPROOFING	9-11
0912.0	**PAINTING (CSI 09900)**	**9-11**
0913.0	**WALL COVERING (CSI 09950)**	**9-12**
.1	VINYL PLASTICS	9-12
.2	PLASTIC WALL TILE	9-12
.3	WALLPAPER	9-12
.4	FABRICS	9-12
.5	LEATHER	9-12
.6	WOOD FLEXIBLE VENEERS	9-12
.7	FIBRE GLASS	9-12
.8	CORK	9-12
.9	GLASS	9-12
.10	GRASS CLOTH	9-12

All items in this division are subcontractors' costs and include all labor, materials, equipment and fees.

0901.0 LATH & PLASTER (L&M) (Latherers and Plaster)

		UNIT	COST
.1	LATHING		
.11	Metal Lath	SqYd	13.00
.12	Gypsum Lath	SqYd	11.50
.13	Metal Studs	SqYd	14.50
.14	Channels - Studs - 16" O.C.	SqYd	14.00
	Furring - Beams & Columns - 16" O.C.	SqYd	17.60
	Ceilings (including hangers) - 16" O.C.	SqYd	19.70
.15	Metal Trim - Base Bead and Corner Bead	LnFt	3.30
	Picture Mould	LnFt	4.40
.2	PLASTERING		
.21	Gypsum (Putty Coat, Sand Float Textured),Walls - 1 Coat	SqYd	15.00
	3 Coat	SqYd	42.00
	Add for Ceiling, Column and Beam Applications	SqYd	4.00
.22	Acoustical	SqYd	24.00
.23	Vermiculite	SqYd	22.00
.24	Stucco or Cement Plaster		
	On Masonry - Large Areas - Float Finish - 1 Coat	SqYd	25.00
	3 Coat	SqYd	33.00
	Panels, Facias, Soffits - 3 Coat	SqYd	38.50
	On Metal Lath - Large Areas - 3 Coat	SqYd	44.00
	Panels, Facias, Soffits, Columns	SqYd	51.00
	On Metal Studs - 4" - 16 ga	SqYd	30.50
	6" - 16 ga	SqYd	42.00
	Deduct for 18 ga Studs	SqYd	10%
	Add for Trowel Finish	SqYd	4.30
.25	Fireproofing (Sprayed On) - Mineral Fiber		
	Steel and Decks - 1 Hour	SqYd	9.30
	2 Hour	SqYd	15.50
	Beams and Columns - Not Wrapped - 2 Hour	LnFt	16.50
	3 Hour	LnFt	19.20
.26	Thin Coat or Veneer Plaster - Walls - 1 Coat	SqYd	13.00
	2 Coat	SqYd	19.80
	Ceilings - 2 Coat	SqYd	25.00
	Add for Patching	SqYd	50%
	Add Color to Plaster and Stucco	SqYd	4.80
	Add for Scaffold - Average	SqYd	8.50
.27	Exterior Wall System - Stucco or Synthetic Plaster		
	Includes Structural Board and 2" Insulation	SqYd	95.00
	Add for Esthetic Grooving	LnFt	4.50

LATH AND GYPSUM PLASTER IN COMBINATION UNITS:		FINISHED	
WALLS	UNIT	1 Side	2 Side
1 Coat Plaster on Waterproofed Found. Walls	SqYd	15.00	-
2 Coat Plaster on Masonry and Concrete	SqYd	30.00	61.00
on Gypsum Lath	SqYd	37.00	84.00
3 Coat Plaster on Metal Lath	SqYd	46.00	94.00
2 Coat Plaster on Metal Studs & Gypsum Lath	SqYd	62.00	100.00
CEILINGS			
3 Coat Plaster on Metal Lath and Channel	SqYd	88.00	-
for Suspended Ceiling (including hangers)			
3 Coat Plaster and Metal Lath attached to	SqYd	73.00	-
joists, beams, etc.			
FURRED			
Beam - 3 Coat Plaster, Metal Lath, Channel	SqYd	88.00	-
Column & Pipe - 3 Coat Plaster - Metal Lath	SqYd	83.00	-
Cabinet - 3 Coat Plaster, Metal Lath, Chnl.	SqYd	56.00	-
2 Coat Plaster on Gypsum Lath	SqYd	52.00	-

0902.0 GYPSUM DRYWALL (L&M) (Carpenters)
Based on 8' Ceiling Heights & Screwed-on Application

		UNIT	COST
.1	BACKER BOARD		
	3/8"	SqFt	.90
	1/2"	SqFt	.92
	5/8" Fire Rated	SqFt	.93
	Deduct for Nailed-on	SqFt	.08
.2	FINISH BOARD - Walls		
	1/2" - Standard	SqFt	1.50
	Fire Rated	SqFt	1.53
	Moisture Resistant	SqFt	1.60
	Insulating (Foil Backed)	SqFt	1.65
	Combined Fire Rated and Insulating	SqFt	1.70
	5/8" - Standard	SqFt	1.50
	Fire Rated	SqFt	1.52
	Moisture Resistant	SqFt	1.52
	Insulating (Foil Backed)	SqFt	1.55
	Combined Fire Rated and Insulating	SqFt	1.62
	Lead Lined with #4 Lead - 1/16"	SqFt	7.00
	#2 Lead - 1/16"	SqFt	5.50
	Add for Ceiling Work - to Wood	SqFt	.15
	Add for Ceiling Work - to Steel	SqFt	.38
	Add for Adhesive Method	SqFt	.16
	Deduct for Nailed On	SqFt	.08
	Add for Beam and Column Work	SqFt	.65
	Add for Resilient Clip Application	SqFt	.75
	Add for Small Cut Up Areas	SqFt	.55
	Add for Each Floor Added	SqFt	.07
	Add for Work Above 10'	SqFt	.23
	Add for Filling Hollow Metal Frames	Each	65.00
.3	VINYL COATED BOARD - 1/2"	SqFt	1.90
	5/8"	SqFt	2.00
.4	LINER PANELS - 1"	SqFt	2.50

			COST	
		UNIT	25 ga	20 ga
.5	STEEL STUDS (Incl. Top & Bottom Plates)			
	16" O.C. - 1 5/8"	SqFt	1.30	1.50
	2 1/2"	SqFt	1.32	1.52
	3 5/8"	SqFt	1.28	1.48
	6"	SqFt	1.34	1.53
	24" O.C. - 1 5/8"	SqFt	1.00	1.22
	2 1/2"	SqFt	1.04	1.25
	3 5/8"	SqFt	1.05	1.30
	6"	SqFt	1.13	1.40
	Add for Work Over 10' Heights	SqFt	.16	.15
	Add for 16 Ga 6" Galvanized	SqFt	1.55	1.15
.6	FURRING CHANNELS			
	16" O.C. 3/4" x 1 3/8" walls	SqFt	-	1.10
	24" O.C. 3/4" x 1 3/8" walls	SqFt	-	1.00
	Resilient Channels - 16" O.C.	SqFt	-	1.00
	Add for Ceiling Work	SqFt	-	.15
	Add for Beam and Column Work	SqFt	-	1.05
.7	METAL TRIM - Casing Bead	LnFt	-	1.05
	Corner Bead	LnFt	-	.82
.8	TAPING AND SANDING	SqFt	-	.48
.9	TEXTURING	SqFt	-	.55

See 0602.34 for Labor & Material Priced Separately

0902.0 GYPSUM DRYWALL, Cont'd...
COMBINATION UNITS - METAL FRAMING & BOARD

PARTITIONS 3 5/8" - 25 ga Studs - 24" O.C.

	UNIT	COST
1/2" Board One Side	SqFt	2.60
Two Sides	SqFt	3.65
1/2" Fire Rated Board Two Sides	SqFt	3.70
1/2" Fire Rated Board One Side - 2 Layers 2nd Side (1 Hr.)	SqFt	4.60
2 Layers Each Side (2 Hr.)	SqFt	5.50
5/8" Board One Side	SqFt	2.45
Two Sides	SqFt	3.95
5/8" Fire Rated Board Each Side (1 Hr.)	SqFt	4.00
5/8" Fire Rated Board - 2 Layers Each Side (2 Hr.)	SqFt	5.70
3 Layers to Columns & Beams (2 Hr.)	SqFt	4.25
Add for Studs 16" O.C. - 25 Ga	SqFt	.25
Deduct for 1 5/8" Studs - 25 Ga	SqFt	.22
Deduct for 2 5/8" Studs - 25 Ga	SqFt	.12
Add for Studs - 20 Ga and 16" O.C.	SqFt	.27
Add for Studs - 20 Ga and 24" O.C.	SqFt	.25
Add for 6" Studs - 25 Ga and 16" O.C.	SqFt	.12
Add for 6" Studs - 20 Ga and 16" O.C.	SqFt	.19
Add for Sound Deadening	SqFt	.80

PARTITIONS - CAVITY SHAFT WALL -
25 Ga Studs - 24" O.C. - All Board Fire Code C

	UNIT	COST
1" Liner Board One Side - 5/8" Board 2nd Side (1 Hr.)	SqFt	4.80
2 Layers - 1/2" Board 2nd Side (2 Hr.)	SqFt	5.25
3 Layers - 5/8" Board 2nd Side (3 Hr.)	SqFt	6.30
1" Liner Board and 5/8" Board One Side		
2 Layers 1" Liner Board 2nd Side	SqFt	8.50

PARTITIONS - DEMOUNTABLE

	UNIT	COST
2 1/2" Studs 24" O.C.-1/2" Vinyl Coated Bd. 2 Sides	SqFt	4.45
3 5/8" Studs 24" O.C.-1/2" Vinyl Coated Bd. 2 Sides	SqFt	4.80
Add for 16" O.C.	SqFt	.30
Add for 5/8" Vinyl Coated Board 2 Sides	SqFt	.33
Add for Special Colors Baked On	SqFt	25%

FURRED WALLS

	UNIT	COST
16" O.C. 1/2" Board	SqFt	2.40
24" O.C. 1/2" Board	SqFt	2.45
Resilient Channels and 1/2" Board	SqFt	2.40

CEILINGS

	UNIT	COST
To Wood Joists - 1/2" Fire Rated Board (1 Hr.)	SqFt	1.25
To Steel Joists with Furring Channels		
24" O.C. - 1/2" F.R. Board (2 Hr. with 2 1/2" Conc.)	SqFt	2.35
5/8" F.R. Board (1 Hr. with 2" Conc.)	SqFt	2.40
12" O.C. - 5/8" F.R. Board (2 Hr. with 2 1/2" Conc.)	SqFt	2.90

COLUMNS - 14 WF and Up

	UNIT	COST
1/2" F.R. Board w/ Screw Studs at Corners (1 Hr.)	SqFt	3.00
2 Layers - 1/2" F.R. Bd. with Screw Studs @ Corners (2 Hr.)	SqFt	3.50
Add for Corner Bead	LnFt	1.10

See 0602.34 for Labor and Material Priced Separately
See 0703.0 for Insulation Work

		UNIT	COST
0903.0 TILEWORK (L&M) (Tile Setters)			
.1 CERAMIC TILE			
Floors - 1" Square - Thin Set (Unglazed)		SqFt	7.80
- 1" Hex - Thin Set		SqFt	8.50
1" x 2"		SqFt	8.60
2" Square		SqFt	8.00
16" Square		SqFt	8.50
Add for Mortar Setting Bed		SqFt	4.50
Add for Random Colors		SqFt	1.00
Add for Epoxy Joints		SqFt	2.30
Conductive - In Mortar Setting Bed		SqFt	12.00
Walls - 2" x 2" - Thin Set		SqFt	9.60
4 1/4" x 4 1/4"		SqFt	7.30
4 1/4" x 6"		SqFt	7.40
4 1/4" x 8"		SqFt	7.70
6" x 6"		SqFt	8.10
Add for Plaster Setting Bed		SqFt	4.40
Add for Decorator Type		SqFt	6.25
Base - 4 1/4"		LnFt	7.50
6"		LnFt	7.45
Add for Coved Base		LnFt	.90
Accessories - Toilet and Bath Embedded		Each	55.00
.2 QUARRY TILE			
Floors - 4" x 4" - Thin Set		SqFt	7.90
6" x 6"		SqFt	7.70
4" x 8"		SqFt	7.80
Hexagon, Valencas, etc.		SqFt	8.85
Walls - 4" x 4" - Thin Set		SqFt	8.60
6" x 6"		SqFt	8.40
Stairs		SqFt	11.10
Base - 4" x 4" x 1/2"		LnFt	8.20
6" x 6" x 1/2"		LnFt	9.10
Add for Hydroment Joints		SqFt	.85
Add for Epoxy Joints		SqFt	2.20
Add for Non-Slip Tile		SqFt	.50
Add for Mortar Setting Bed		SqFt	4.50
0904.0 TERRAZZO (L&M) (Terrazzo Workers) - Based on Gray Cement, 4' Panel, and White Dividing Strips			
.1 CEMENT TERRAZZO			
Floors - 2" Bonded		SqFt	11.60
2 1/2" Not Bonded		SqFt	10.20
Base - 6" without Grounds or Base Bead		LnFt	18.50
6" with Grounds and Base Bead		LnFt	22.50
Stairs - Treads and Risers - Tread Length		LnFt	47.50
Stringers without Base Bead		LnFt	42.50
Pan Filled Treads and Landings		LnFt	19.60
Add for White Cement		SqFt	.70
Add for Brass Dividing Strips		SqFt	1.00
Add for Large Chips (Venetian) - 3"		SqFt	5.55
Add for 2-Color Panels		SqFt	3.30
Add for 3-Color Panels		SqFt	3.90
Add for Non-Slip Abrasives		SqFt	.96
Add for Small or Scattered Areas		SqFt	15%
.2 CONDUCTIVE TERRAZZO - 2" Bonded		SqFt	12.50
1/4" Thin Set Epoxy		SqFt	11.30
.3 EPOXY TERRAZZO - 1/4" Thin Set		SqFt	9.50
.4 RUSTIC TERRAZZO (WASHED AGGREGATE) - Marble		SqFt	6.40
Quartz		SqFt	7.00

See 0907 for Precast Terrazzo Tiles

0905.0 ACOUSTICAL TREATMENT (L&M) (Carpenters)

		UNIT	COST
.1	MINERAL TILE (No Backing or Supports)		
	Fissured or Perforated - 1/2" x 12" x 12" or 12" x 24"	SqFt	2.00
	5/8" x 12" x 12" or 12" x 24"	SqFt	2.25
	3/4" x 12" x 12" or 12" x 24"	SqFt	2.40
	Add to all above for Backing or Supports if needed:		
	Ceramic Faces	SqFt	.49
	Concealed Z Spline Application	SqFt	.62
	Wood Furring Strips	SqFt	.68
	Rock Lath	SqFt	.90
	1 1/2" Channel Suspension	SqFt	1.25
	Resilient Furring Channels	SqFt	.90
.2	PLASTIC COVERED TILE (No Backing or Supports)		
	3/4" x 12" x 12"	SqFt	2.30
	5/8" x 12" x 12"	SqFt	1.90
.3	METAL PAN TILE (Supports Included)		
	12" x 24" or 12" x 36" or 12" x 48" - Steel (24 Ga)	SqFt	6.90
	Aluminum (.025)	SqFt	8.50
	Aluminum (.032)	SqFt	9.80
	Stainless Steel	SqFt	11.70
	Add for Color	SqFt	.50
.4	CLASS A MINERAL TILE		
	(Grid System w/ Suspension incl. & Partitions to Clg.)		
	5/8" x 24" x 24" or 24" x 48" - Perforated	SqFt	1.65
	Fissured	SqFt	1.65
	Add for Partitions thru Ceilings	SqFt	.25
	Add for Translucent Panels for Lights	SqFt	1.35
.5	FIRE RATED DESIGN MINERAL TILE		
	(Grid System w/ Suspension incl. & Partitions to Clg.)		
	5/8" x 24" x 24" or 24" x 48" - Perforated (2-hour rating)	SqFt	1.80
	Fissured (2-hour rating)	SqFt	1.80
	3/4" x 24" x 24" or 24" x 48" - Perforated (3-4 Hr rating)	SqFt	1.95
	Fissured (3-4 Hr rating)	SqFt	1.95
	Add for Partitions thru Ceilings	SqFt	.25
	Add for Reveal Edge or Shadow Line (24" x 24" only)	SqFt	.35
	Add for Nubby Tile	SqFt	1.25
	Add for Ceramic Faces (5/8")	SqFt	.75
	Add for Colored Tile	SqFt	.25
	Add for Fixture Fire Protection - Fixtures only	SqFt	1.05
.6	GLASS FIBRE BOARD (Grid System w/ Suspension included)		
	5/8" x 24" x 24"	SqFt	1.75
	3/4" x 24" x 48"	SqFt	1.80
	1" x 48" x 48"	SqFt	1.75
	1½" x 60" x 60"	SqFt	3.10
	Add for Three-Dimensional	SqFt	3.00
.7	SOUND CONTROL PANELS AND BAFFLES	SqFt	12.30

		UNIT	COST		
.8	ACOUSTICAL WALL PANELS - Special				
	Fiberglass with Cloth Face - 3/4"	SqFt	4.00	to	10.00
	1"	SqFt	5.00	to	11.50
	1½"	SqFt	8.00	to	13.50
	3/4" Mineral	SqFt	1.30	to	3.60
	1/2" Fibre Board	SqFt	.85	to	1.25
.9	LINEAR METAL CEILINGS				
	4" Wide - Painted	SqFt	5.80	to	7.60
	Chrome or Brass	SqFt	8.30	to	10.50

See 0307 for Fibre Cementitious Acoustical Plank

0906.0 VENEER STONE (Floors, Walls, Stools, Stairs, Partitions) (2" and Under) (L&M) (Marble Setters)

		UNIT	COST
.1	LIMESTONE	SqFt	29.00
.11	Cut Stone - Floors 7/8" Interior	SqFt	29.00
	Floors 2" Exterior	SqFt	33.00
	Walls 1"	SqFt	32.00
	Stools 1" x 8"	LnFt	39.00
.12	Flagstone Floors 1"	SqFt	35.00
.2	MARBLE Floors 5/8"	SqFt	30.00
	Floor Tiles 3/8" x 12" x 12"	SqFt	32.00
	Walls 3/8" x 12" x 12"	SqFt	35.00
	Base 1" x 6"	LnFt	26.00
	Stool 1" x 8"	LnFt	36.00
	Stairs - Treaded Risers	SqFt	70.00
.3	SLATE Floors 3/8" Package	SqFt	19.50
	Floors 3/4" Random or Flagstone	SqFt	21.00
	Walls 3/4" Random	SqFt	26.00
	Stools 1" x 8"	LnFt	37.00
.4	SANDSTONE Floors 2"	SqFt	33.00
.5	GRANITE Floors 5/6" Interior	SqFt	39.00
	Walls - 2" Exterior	SqFt	47.00
	1 1/2" Interior	SqFt	43.00
	Counter Tops 1 1/4"	SqFt	64.00
.6	SYNTHETIC MARBLE Floors 5/8"	SqFt	17.50
	Walls 5/8"	SqFt	20.00
	Add for Leveling Old Floors	SqFt	1.75 to 4.00

0907.0 SPECIAL FLOORING (L&M)

		UNIT	COST
.1	MAGNESIUM OXYCHLORIDE (Seamless)	SqFt	8.30
.2	ELASTOMERIC	SqFt	9.50
.3	SYNTHETIC		
	Epoxy with Granules 1/16" Single Speed	SqFt	4.80
	3/32" Double Speed	SqFt	5.80
	Epoxy with Aggregates 1/4" Industrial	SqFt	7.50
	Polyester with Granules 3/32"	SqFt	6.30
	1/4"	SqFt	7.80
	See 0904.3 for Epoxy Terrazzo and 0603.2 for Counter Tops		
.4	ASPHALTIC MASTIC	SqFt	7.60
	Add for Acid Proof	SqFt	1.35
.5	BRICK PAVERS - Mortar Setting Bed Application		
	3 7/8" x 8" x 1 1/4" - Red	SqFt	9.25
	Iron Spot	SqFt	10.50
	Acid Proof	SqFt	20.00
	See 0206.3 for Patios and Walks on Sand Cushion		
.6	PRECAST PANELS		
	12" x 12" x 1" Precast Terrazzo	SqFt	25.00
	12" x 12" x 1" Precast Exposed Aggregate	SqFt	25.00
	12" x 12" x 1" Ceramic (Terra Cotta)	SqFt	30.00
	6" Precast Terrazzo Base	LnFt	20.00
.7	NEOPRENE 1/4"	SqFt	11.00

0908.0 RESILIENT FLOORING (L&M)
(Carpet and Resilient Floor Layers)

		UNIT	COST
.1	ASPHALT - Tile 9" x 9" x 1/8"	SqFt	2.40
.2	CORK - Tile 1/8" - Many Sizes	SqFt	3.15
	3/16" - Many Sizes	SqFt	3.70
.3	RUBBER		
	Tile 12" x 12" x 3/32"	SqFt	4.40
	12" x 12" x 3/32" Conductive	SqFt	5.50
	Base 2 1/2"	LnFt	1.35
	4"	LnFt	1.50
	6"	LnFt	1.70
	Treads, Stringers and Risers		
	Treads 3/16" Gauge Pre-formed	LnFt	10.50
	1/4" Gauge Pre-formed	LnFt	12.00
	5/16" Gauge Pre-formed	LnFt	15.50
	Stringers 10" x 1/8"	LnFt	8.00
	Risers	LnFt	4.20
.4	VINYL COMPOSITION TILE 12" x 12" x 3/32"	SqFt	1.45
	12" x 12" x 1/8"	SqFt	1.55
	Conductive	SqFt	5.25
	Treads 1/8"	LnFt	5.60
	3/16"	LnFt	7.00
	1/4"	LnFt	8.75
	Stringers 10" x 1/8"	LnFt	8.65
	Risers 7" x 1/8"	LnFt	3.90
	Add for Marbleized or Patterns	SqFt	1.20
.5	VINYL TILE 9" x 9" x 1/8" Solid	SqFt	6.20
	12" x 12" x 1/8"	SqFt	6.40
	Base 2 1/2"	LnFt	1.50
	4"	LnFt	1.55
	6"	LnFt	1.75
.6	VINYL SHEET 1/8" Inlaid	SqFt	2.45 - 6.70
	1/8" Homogenous	SqFt	4.10 - 7.20
	Acoustic Cushioned	SqFt	4.50 - 10.40
	Coving	LnFt	6.00
.7	VINYL CORNER GUARDS	LnFt	5.30
	Add Plywood or Masonite Underlayment	SqFt	1.60
	Add Diagonal Cuttings	SqFt	.55
	Add Clean and Wax	SqFt	.42
	Add Skim Coat Leveling	SqFt	1.85 - 3.40

0909.0 CARPETING (L&M) (Carpet Layers)

		UNIT	CUT PILE	LOOP PILE	
.1	NYLON - 20 oz	SqYd		13.50	- 21.50
	22 oz	SqYd		14.50	- 22.50
	24 oz	SqYd	12.00 - 16.00	15.50	- 26.00
	26 oz	SqYd		16.50	- 26.00
	28 oz	SqYd		16.50	- 27.00
	30 oz	SqYd	14.50 - 20.00	18.00	- 30.00
	32 oz	SqYd	15.50 - 21.00	19.00	- 32.00
.2	WOOL - 36 oz	SqYd		25.00	- 60.00
	42 oz	SqYd		32.00	- 72.00
.3	OLEFIN - 20 oz	SqYd		12.00	- 19.00
	22 oz	SqYd		13.00	- 20.00
	24 oz	SqYd		14.50	- 21.00
	26 oz	SqYd		15.50	- 24.00
	28 oz	SqYd		17.00	- 25.00
.4	CARPET TILE, MODULAR-Tension Bonded	SqYd		34.00	- 60.00
	Tufted	SqYd	23.00 - 36.00	23.00	- 44.00
.5	INDOOR-OUTDOOR	SqYd		9.50	- 19.00

Installation included in above prices   3.50 - 6.00

0909.0	CARPETING, Cont'd...	UNIT	COST
	Add to Section 0909.0 above:		
	Add for Custom Carpet	SqYd	9.00 - 19.00
	Add for Cushion (Urethane, Rubber, Synthetic)	SqYd	3.70 - 6.00
	Add Bonded Cushion (Urethane Backed)	SqYd	1.30 - 1.50
	Add for Carpet Base, Top Edge Bound	LnFt	1.90 - 3.00
	Add for Skim Coat Leveling, Latex	SqFt	1.10 - 1.50
	Add for Border and Inset Work	LnFt	2.30 - 3.70
	Add for Reduction Strips	LnFt	1.75 - 2.60

0910.0 WOOD FLOORING (L&M) (Carpenters)

				UNIT	COST
.1	STRIP AND PLANK FLOORING				
	Oak Floors (based on Standard Grade)				
	White Oak 25/32 1 1/2" Select Job Finished			SqFt	9.50
	2 1/4" Select Job Finished			SqFt	9.70
	3 1/4" Select Job Finished			SqFt	10.00
	Add for Clear Grade			SqFt	.35
	Deduct for Antique or Rustic Floors			SqFt	.45
	Maple Floors - Iron Bound (Edge Grain)				
	3/4" x 1 1/2"	#2 & Better Gr	Job Finished	SqFt	10.20
	3/4" x 2 1/4"	#2 & Better Gr	Job Finished	SqFt	10.75
	33/32 x 2 1/4"	#2 & Better Gr	Job Finished	SqFt	10.20
	33/32 x 2 1/4"	#3 Grade	Job Finished	SqFt	9.50
	Add for #1 Grade			SqFt	1.10
	Add to above for Cork Underlayment			SqFt	1.20
	Fir Floors				
	1" x 4" Vertical Grain B&B Job Finished			SqFt	5.00
	Flat Grain B&B Job Finished			SqFt	4.90
.2	WOOD PARQUET Pre Finished 5/16" Oak or Maple			SqFt	7.75
	FLOORING Pre Finished 3/4 " Oak or Maple			SqFt	9.70
	Pre Finished 5/16" Teak			SqFt	10.50
	5/16" Walnut			SqFt	10.80
.3	RADIANT TREATED (Plastic and Wood) 5/16" x 12" x 12"			SqFt	11.70
.4	RESILIENT WOOD FLOOR SYSTEMS				
	Cushion Sleeper Floors (Maple)				
	33/32 2 1/4" #1 Grade Pre-Finished			SqFt	8.80
	#2 Grade Pre-Finished			SqFt	8.40
	#3 Grade Pre-Finished			SqFt	7.80
	Cork Underlayment included			SqFt	1.25
	Add for Plywood Underlayment (2 layers)			SqFt	1.90
	Fixed Sleeper Floors (Oak)				
	33/32 2 1/4" #1 Grade Pre-Finished			SqFt	8.80
	#2 Grade Pre-Finished			SqFt	8.65
	1/4" #3 Grade Pre-Finished			SqFt	8.55
.5	WOOD BLOCK INDUSTRIAL FLOORS (Creosoted) 2"			SqFt	4.10
	2 1/2"			SqFt	4.35
	3"			SqFt	4.50
.6	FINISHING (Sanding, Sealing and Waxing)				
	Included in Job Finished Prices Above for New House			SqFt	1.80
	for New Gym			SqFt	1.75
	for Old Gym			SqFt	2.90
	Add to Section 0910.0 above:				
	Small Cut-up Areas			SqFt	30%
	Trim and Irregular Areas			SqFt	15%
	Polyurethane Finish with Stain			SqFt	20%
	Deduct for Residential Work			SqFt	10%

0911.0 SPECIAL COATINGS

	UNIT	COST
.1 CEMENTITIOUS (Glazed Cement)	SqFt	2.35 - 4.00
.2 EPOXIES	SqFt	1.45 - 2.75
.3 POLYESTERS	SqFt	1.50 - 2.85
.4 VINYL CHLORIDE	SqFt	1.70 - 3.10
.5 AGGREGATE TO EPOXY - Blown on Quartz	SqFt	7.35
Troweled on Quartz	SqFt	7.60

	UNIT	COST
.6 SPRAYED FIREPROOFING (with Perlite)		
Column - 1" 2 Hour	LnFt	1.50
1 3/8" 3 Hour	LnFt	1.80
1 3/4" 4 Hour	LnFt	2.30
Beams - 1 1/8" 2 Hour	LnFt	1.65
1 1/4" 3 Hour	LnFt	1.70
1 1/2" 4 Hour	LnFt	2.00
Ceilings - 1/2" 1 Hour	SqFt	1.25
5/8" 2 Hour	SqFt	1.35
1" 3 Hour	SqFt	1.60

See 0901.25 for Other Fireproofing

0912.0 PAINTING (L&M) (Painters)

WALLS AND CEILINGS - BRUSH

	UNIT	COST
Brick 2 Coats	SqFt	.70
Concrete 2 Coats	SqFt	.63
Concrete Block - Interior 2 Coats	SqFt	.66
Exterior 2 Coats	SqFt	.68
Light Weight 2 Coats	SqFt	.71
2 Coats/Filler	SqFt	.87
Plaster 2 Coats	SqFt	.65
Sheet Rock 2 Coats	SqFt	.56
Steel Decking & Sidings 1 Coat	SqFt	.47
Stucco 2 Coats	SqFt	.78
Wood Siding & Shingles 2 Coats	SqFt	.77
Wood Paneling - Stain, Seal and Varnish	SqFt	1.45
Deduct for Roller Application	SqFt	.11
Deduct for Spray Application	SqFt	.15
Add for Scaffold Work	SqFt	.23
Add for Epoxy Paint	SqFt	.21

DOORS AND WINDOWS

	UNIT	COST
Door, Wood - 2 Coats	Each	39.00
Stain, Seal and Varnish	Each	59.00
Bi-fold Door - Stain, Seal and Varnish 48"	Each	56.00
60"	Each	66.00
72"	Each	78.00
Door Frame, Wood (incl Trim) - 2 Coats	Each	35.00
Stain, Seal & Varnish	Each	43.00
Door, Metal (Factory Primed) - 1 Coat	Each	32.00
2 Coats	Each	43.00
Door Frame, Metal (Factory Primed) - 1 Coat	Each	29.00
2 Coats	Each	32.00
Window, Wood (including Frame & Trim) - 2 Coats	Each	43.00
Add for Storm Windows or Screens	Each	22.00
Add for Shutters - Pair	Pair	63.00
Window, Steel	Each	40.00

	UNIT	COST
FLOORS, EPOXY	SqFt	1.95

		UNIT	COST
0912.0	**PAINTING (L&M) (Painters)**		
	MISCELLANEOUS WOOD		
	Cabinets - Enamel 2 Coats	SqFt	.77
	Stain and Varnish	SqFt	.96
	Enamel 3 Coats	SqFt	1.15
	Stain, Seal and Varnish	SqFt	1.50
	Wood Trim - Stain, Seal and Varnish - Large	LnFt	1.50
	Small or Decorative	LnFt	1.75
	MISCELLANEOUS METAL - 1 Coat Sprayed		
	Miscellaneous Iron and Small Steel	Ton	.96
	Structural Steel	Ton	.63
	Structural Steel - 2 Sq.Ft. per Ln.Ft.	LnFt	.48
	3 Sq.Ft. per Ln.Ft.	LnFt	.60
	4 Sq.Ft. per Ln.Ft.	LnFt	.82
	5 Sq.Ft. per Ln.Ft.	LnFt	1.02
	Add for Brush	LnFt	40%
	Add for 2 Coats	LnFt	75%
	Gutters, Down Spouts and Flashings - 4"	LnFt	1.30
	5"	LnFt	1.35
	6"	LnFt	1.50
	Piping - Prime and 1 Coat - 4"	LnFt	.65
	6"	LnFt	.92
	8"	LnFt	1.05
	10"	LnFt	1.40
	Radiators	SqFt	2.80
	MISCELLANEOUS		
	Gold and Silver Leaf	SqFt	42.00
	Parking Lines - Striping	LnFt	.30
	Wood Preservatives	SqFt	.52
	Waterproof Paints - 2 Coats	SqFt	.90
	Clear - 1 Coat	SqFt	.47
	2 Coat	SqFt	.67
	Taping and Sanding	SqFt	.45
	Texturing	SqFt	.50

See 0910 for Wood Floor Finishing

		UNIT	COST	
0913.0	**WALL COVERINGS (L&M) (Painters)**			
.1	VINYL PLASTICS (54" Roll)			
	Light Weight	SqFt	1.45 -	1.80
	Medium Weight	SqFt	1.50 -	2.05
	Heavy Weight	SqFt	1.60 -	3.70
	Add for Sizing	SqFt	.18 -	.30
	Add for 27" and 36" Rolls	SqFt	.15 -	.33
.2	PLASTIC WALL TILE	SqFt	2.75 -	3.40
.3	WALLPAPER ($10.00-40.00 Roll, 36 S.F. to Roll)	SqFt	1.40 -	3.15
	Add or Deduct per Roll - $1.00	SqFt	.17 -	.70
.4	FABRICS (Canvas, etc.)	SqFt	1.50 -	3.10
.5	LEATHER	SqFt	2.00 -	3.20
.6	WOOD VENEERS	SqFt	5.10 -	9.20
.7	FIBERGLASS	SqFt	2.10 -	3.45
.8	CORK - 3/16"	SqFt	1.75 -	3.80
.9	GLASS - 5/16"	SqFt	1.15 -	2.05
.10	GRASS CLOTH	SqFt	1.65 -	2.80

		PAGE
1001.0	ACCESS PANELS	10-2
1002.0	CHALKBOARDS AND TACKBOARDS (CSI 10110)	10-2
1003.0	CHUTES (Laundry and Waste) (CSI 11175)	10-2
1004.0	COAT & HAT RACKS AND CLOSET ACCESSORIES (CSI 10914 & 10916)	10-2
1005.0	CUBICLE CURTAIN TRACK & CURTAINS (CSI 10199)	10-2
1006.0	DIRECTORIES AND BULLETIN BOARDS (CSI 10410)	10-3
1007.0	DISPLAY AND TROPHY CASES (CSI 10110)	10-3
1008.0	FLOORS - ACCESS OR PEDESTAL (CSI 10270)	10-3
1009.0	FIREPLACES AND STOVES (CSI 10300)	10-3
1010.0	FIRE FIGHTING SPECIALTIES (CSI 10520)	10-3
1011.0	FLAG POLES (CSI 10350)	10-3
1012.0	FLOOR AND ROOF ACCESS DOORS (CSI 07720)	10-4
1013.0	FLOOR MATS AND FRAMES (CSI 12690)	10-4
1014.0	FOLDING GATES (CSI 10610)	10-4
1015.0	LOCKERS (CSI 10500)	10-4
1016.0	LOUVERS AND VENTS (CSI 10240 & 10235)	10-5
1017.0	PARTITIONS - WIRE MESH (CSI 10605)	10-5
1018.0	PARTITIONS AND CUBICLES - DEMOUNTABLE (CSI 10610)	10-5
1019.0	PARTITIONS - OPERABLE (CSI 10650)	10-5
1020.0	PEDESTRIAN CONTROLS - GATES & TURNSTILES (CSI 10450)	10-6
1021.0	POSTAL SPECIALTIES (CSI 10550)	10-6
1022.0	SHOWER & TUB RECEPTORS AND DOORS (CSI 1082)	10-6
1023.0	SIGNS, LETTERS AND PLAQUES (CSI 10420 & 10480)	10-6
1024.0	STORAGE SHELVING (CSI 10670)	10-6
1025.0	SUN CONTROL AND PROTECTIVE COVERS (CSI 10530)	10-6
1026.0	TELEPHONE AND SOUND BOOTHS (CSI 10570)	10-6
1027.0	TOILET AND BATH ACCESSORIES (CSI 10800)	10-6
1028.0	TOILET COMPARTMENTS (Including Shower and Dressing Rooms) (CSI 10155)	10-7
1029.0	URNS AND TRAYS (CSI 10800)	10-7
1030.0	WALL AND CORNER GUARDS (CSI 10260)	10-7

		UNIT	LABOR	MATERIAL Flush Flange	MATERIAL Drywall	MATERIAL UL Label
1001.0	**ACCESS PANELS (M) (Lathers, Carpenters)**					
	10" x 10"	Each	14.00	32.00	35.00	105.00
	12" x 12"	Each	15.00	37.00	-	110.00
	14" x 14"	Each	16.00	40.00	40.00	130.00
	16" x 16"	Each	17.00	42.00	-	145.00
	18" x 18"	Each	19.00	50.00	-	150.00
	22" x 22"	Each	21.00	65.00	48.00	175.00
	24" x 24"	Each	24.00	78.00	-	195.00
	24" x 36"	Each	28.00	90.00	-	240.00
	32" x 32"	Each	32.00	105.00	-	260.00
	Add for Plaster Type					15%
	Add for Locks					8.00

		UNIT	LABOR	MATERIAL
1002.0	**CHALKBOARDS AND TACKBOARDS (Carpenters)**			
	Chalkboards			
	Slate - Natural 3/8" - New	SqFt	3.40	15.50
	Resurfaced	SqFt	3.40	8.00
	Porcelain Enameled Steel			
	1/2" Foil Backed Gypsum and 24 Ga	SqFt	2.90	7.50
	28 Ga	SqFt	2.95	7.00
	Marker (White) Board	SqFt	2.80	10.00
	Tackboard			
	Cork - 1/4" on 1/4" Gypsum Backing - Washable	SqFt	1.50	6.00
	Add for Colored Cork	SqFt	-	2.70
	Deduct for Natural Style - Not Washable	SqFt	-	2.50
	Trim (Aluminum)			
	Trim-Slip & Snap (incl. Metal Grounds) 1/4#	LnFt	2.25	1.85
	1/2#	LnFt	2.50	3.60
	Map Rail - 3/4 lb/ft	LnFt	2.85	5.00
	1/4 lb/ft	LnFt	2.50	2.80
	Chalk Tray	LnFt	3.40	6.10
	Average Trim Cost	SqFt	2.60	5.40
	Combination Unit Costs (Average)			
	Slate - 3/8" - with Tackboard and Trim	SqFt	3.60	12.80
	Enameled Steel (24 Ga) w/Tackbd. & Trim, 8' x 4'	SqFt	3.40	7.75
	Marker (White) Board	SqFt	3.50	8.00
	Tackboard with Trim	SqFt	3.60	10.00
	Add Vertical and Horizontal Sliding Units	SqFt	-	5.00
	Portable Chalkboard Tackboard - 4' x 6'	Each	60.00	800.00
	Add for Revolving Type		-	270.00
1003.0	**CHUTES (Laundry & Rubbish)(M)(Sheet Metal Workers)**			
	Laundry or Linen			
	(16 Ga Aluminized Steel and 10' Floor Height)			
	1 Floor Opening - 24" Dia - per Opening	Opng	200.00	860.00
	2 Floors or Openings	Opng	190.00	710.00
	6 Floors or Openings	Opng	180.00	660.00
	Average	LnFt	20.00	75.00
	Rubbish			
	(16 Ga Aluminized Steel)			
	1 Floor Opening - 24" Dia - per Opening	Opng	200.00	860.00
	2 Floors or Openings	Opng	190.00	730.00
	6 Floors or Openings	Opng	175.00	640.00
	Add for Explosion Vent - per Opening	Opng	-	425.00
	Add for Stainless Steel	Opng	-	360.00
	Add for 30" Diameter	Opng	15.00	170.00
	Deduct for Galvanized	Opng	-	75.00
1004.0	**COAT/ HAT RACKS & CLOSET ACCESS. (M) (Carpenters)**			
	ALUMINUM	LnFt	12.00	70.00
	STAINLESS STEEL	LnFt	12.00	82.00
1005.0	**CUBICLE CURTAIN TRACK & CURTAINS (M) (Carpenters)**	LnFt	8.00	15.00

1006.0 DIRECTORIES & BULLETIN BOARDS (M) (Carpenters)
(Directories For Changeable Letters)

	UNIT	LABOR	MATERIAL
Interior - 48" x 36" Aluminum	Each	130.00	880.00
Stainless Steel	Each	135.00	980.00
Bronze	Each	135.00	1,350.00
60" x 36" Aluminum	Each	150.00	1,100.00
Stainless Steel	Each	155.00	1,480.00
Bronze	Each	155.00	2,150.00
Exterior - 48" x 36" Aluminum	Each	175.00	1,250.00
Stainless Steel	Each	185.00	1,740.00
Bronze	Each	185.00	2,350.00
60" x 36" Aluminum	Each	185.00	1,730.00
Stainless Steel	Each	200.00	2,240.00
Bronze	Each	200.00	2,900.00
Illuminated - 72" x 48" Aluminum	Each	200.00	5,500.00
Stainless Steel	Each	300.00	8,100.00
Bronze	Each	300.00	9,700.00

1007.0 DISPLAY & TROPHY CASES (M) (Carpenters)

	UNIT	LABOR	MATERIAL
Aluminum Frame 8' x 6' x 1'6"	Each	300.00	3,200.00
4' x 6' x 1'6"	Each	180.00	1,775.00

1008.0 FLOORS - ACCESS OR PEDESTAL (L&M) (Carpenters)
(Includes Laminate Topping)

	UNIT	LABOR	MATERIAL
Aluminum	SqFt		18.50
Steel - Galvanized	SqFt		12.00
Plywood - Metal Covered	SqFt		11.00
Add for Carpeting	SqFt		3.00
Add for Stairs and Ramps	SqFt		-

1009.0 FIREPLACE AND STOVES

	UNIT	LABOR	MATERIAL
	-	-	-

1010.0 FIRE FIGHTING DEVICES (M) (Carpenters)

	UNIT	LABOR	MATERIAL
Extinguishers - 2½ Gal - Water Pressure	Each	22.00	74.00
Carbon Dioxide - 5 Lb	Each	23.00	170.00
10 Lb	Each	24.00	225.00
5 Lb	Each	22.00	53.00
Dry Chemical - 10 Lb	Each	23.00	65.00
20 Lb	Each	26.00	108.00
Halon - 2½ Lb	Each	22.00	115.00
5 Lb	Each	23.00	140.00
Cabinets - Enameled Steel - 18 Ga			
12" x 27" x 8" - Recessed - Portable Ext	Each	40.00	64.00
24" x 40" x 8" - Recessed - Hose Rack w/Nozzle	Each	75.00	135.00
Deduct for Surface Plastic	Each	-	20.00
Add for Aluminum	Each	-	60.00
Blanket & Cabinet - Enameled and Roll Type	Each	40.00	75.00

1011.0 FLAG POLES (M) (Carpenters)

	UNIT	LABOR	MATERIAL
Aluminum Tapered - Ground 20' - 5" Butt		170.00	950.00
25' - 5½" Butt		190.00	1,330.00
30' - 6" Butt		235.00	1,450.00
35' - 7" Butt		310.00	1,920.00
40' - 8" Butt		420.00	3,150.00
50' - 10" Butt		525.00	4,200.00
60' - 12" Butt		670.00	7,400.00
Wall 15' - 4" Butt		180.00	880.00
Add for Architectural Grade			25%
Fiberglass Tapered - Deduct from Above Prices			20%
Steel Tapered - Deduct from Above Prices			20%
Steel Tube - Deduct from Above Prices			40%
Add for Concrete Bases		300.00	200.00

1012.0 FLOOR & ROOF ACCESS DOORS (M) (Carpenters)

	UNIT	LABOR	MATERIAL
Floor - Single Leaf 1/4" Steel			
24" x 24"	Each	85.00	525.00
30" x 30"	Each	110.00	560.00
30" x 36"	Each	135.00	640.00
36" x 36"	Each	160.00	670.00
Deduct for Recess Type	Each		30.00
Add for Aluminum	Each		80.00
Roof - Steel			
30" x 36"	Each	140.00	485.00
30" x 54"	Each	150.00	680.00
30" x 96"	Each	220.00	1,260.00
Add for Aluminum	Each		80.00
Add for Fusible Link	Each		30.00

1013.0 FLOOR MATS & FRAMES (M) (Carpenters)

	UNIT	LABOR	MATERIAL
Mats - Vinyl Link 4' x 5' x 1/2"	Each	37.00	320.00
4' x 7' x 1/2"	Each	40.00	465.00
Add for Color	SqFt		4.00
Frames - Aluminum 4' x 5'	Each	70.00	200.00
4' x 7'	Each	75.00	250.00

1014.0 FOLDING GATES (M) (Carpenters)

	UNIT	LABOR	MATERIAL
8' x 8' - with Lock Cylinder	Each	270.00	1,280.00
	or SqFt	4.25	20.00

1015.0 LOCKERS (Incl. Benches) (M) (Sheet Metal Workers)

	UNIT	LABOR	MATERIAL 60"	72"
Single - Open Type with 6" Legs				
9" x 12" x 72"	Each	19.00	105.00	110.00
9" x 15" x 72"	Each	19.00	110.00	115.00
12" x 12" x 72"	Each	20.00	122.00	125.00
12" x 15" x 72"	Each	21.00	123.00	127.00
12" x 18" x 72"	Each	21.50	125.00	128.00
15" x 15" x 72"	Each	22.50	127.00	131.00
15" x 18" x 72"	Each	25.00	130.00	135.00
15" x 21" x 72"	Each	26.00	140.00	145.00
Double Tier				
12" x 12" x 36"	Each	14.50	65.00	
12" x 15" x 36"	Each	15.00	68.00	
12" x 18" x 36"	Each	15.50	70.00	
15" x 15" x 36"	Each	15.50	73.00	
15" x 18" x 36"	Each	16.00	80.00	
15" x 21" x 36"	Each	16.50	85.00	
Box Lockers				
12" x 12" x 12"	Each	13.50	37.00	
15" x 15" x 12"	Each	14.00	38.00	
18" x 18" x 12"	Each	14.50	40.00	
12" x 12" x 24"	Each	14.50	50.00	
15" x 15" x 24"	Each	15.00	58.00	
18" x 18" x 24"	Each	15.50	60.00	
Add Flat Key Locks (Master Keyed)	Each	-	8.00	
Add for Combination Locks	Each	-	10.00	
Add for Sloped Tops	Each	7.00	10.00	
Add for Closed Base	Each	7.00	8.00	
Add for Other than Standard Colors	Each	-	15%	
Add for Benches	LnFt	8.00	25.00	
Add for Less than Banks of 4	Each	5%	10%	
Add for Recessed Latch (Quiet)	Each	-	3.00	

1016.0 LOUVERS AND VENTS (M) (Carpenters or Sheet Metal Workers)

LOUVRES	UNIT	LABOR	MATERIAL
Exterior			
Galvanized - 16 Ga	SqFt	13.00	36.00
Aluminum - 14 Ga	SqFt	13.00	47.00
Extruded 12 Ga	SqFt	13.00	54.00
Add for Brass, Bronze or Stainless	SqFt	-	42.00
Add for Insect Screen	SqFt	-	3.00
Interior Door or Partition - Steel	SqFt	12.00	42.00
Aluminum	SqFt	12.00	52.00
VENTS - Brick - Aluminum	Each	24.00	65.00
Block - Aluminum	Each	24.00	100.00
See Div. 15 for Mechanically Connected			

1017.0 PARTITIONS - WIRE MESH (M) (Carpenters)

PARTITIONS			
6 Ga - 2" Mesh & Channel Frame	SqFt	4.50	9.30
10 Ga - 1 1/2" Mesh - 8'	SqFt	3.50	7.50
Add for Door	Each	90.00	360.00
WINDOW GUARDS - 10 Ga with Frame	SqFt	5.00	15.00
Add for Galvanized	-	-	15%

1018.0 PARTITIONS AND CUBICLES - DEMOUNTABLE (L&M) (Carpenters)

PARTITIONS - DEMOUNTABLE - 1 9/16" x 10'	UNIT	COST
Aluminum or Enameled Steel Extrusions (Acoustical)		
Gypsum Panels (Unpainted)	SqFt	11.80
Vinyl Laminated Gypsum Panels	SqFt	13.40
Steel Panels	SqFt	13.40
Wood Panels - Hard Wood	SqFt	8.50
Add for 2 3/4"	SqFt	1.50
Add per Door	Each	460.00
CUBICLES - DEMOUNTABLE (Includes Glass)		
Steel Frame and Panels (Acoustical) Burlap		
3' - 6" High	SqFt	21.00
4' - 8" High	SqFt	19.00
5' - 8" High	SqFt	16.50
7' - 0" High	SqFt	17.00
Aluminum Frame and Panels (Acoustical) Burlap		
3' - 6" High	SqFt	17.00
4' - 8" High	SqFt	16.00
5' - 8" High	SqFt	15.50
7' - 0" High	SqFt	15.00
Deduct for Non-Acoustical Type	SqFt	2.00
Add for Vinyl Laminated	SqFt	1.75
Add for Carpeted	SqFt	4.00
Add for Curved Panels	SqFt	7.00
Add per Door	Each	400.00
Add per Gate	Each	250.00

1019.0 PARTITIONS - OPERABLE (Including Moving Dividers and Operable Walls) (L&M) (Carpenters)

	UNIT	LABOR	MATERIAL
ACCORDION FOLD (Fabric)	SqFt	4.50	14.00
Add for Acoustical Type	SqFt	-	4.90
Add for Metal & Wood Supports	LnFt	4.25	5.80
FOLDING (Wood, Plastic Lam., Fabric Cover)	SqFt	5.50	29.00
Add for Acoustical Type	SqFt	-	5.50

1019.0 PARTITIONS - OPERABLE, Cont'd...	UNIT	LABOR	MATERIAL
OPERABLE WALLS (Electrically Operated)			
Folding - Plastic Laminated	SqFt	6.50	37.00
Vinyl Overlaid	SqFt	6.50	35.00
Wood (Prefinished)	SqFt	6.00	34.00
Add for Chalk Bond	-	-	-
Side Coiling - Wood	SqFt	-	45.00
Metal	SqFt	-	50.00
REMOVABLE OR PORTABLE WALLS (Acoustical)			
4"	SqFt	-	50.00
2 1/4"	SqFt	-	40.00
1020.0 PEDESTRIAN CONTROLS - GATES & TURNSTILES	-	-	-
1021.0 POSTAL SPECIALTIES (M) (Carpenters)			
CHUTES - Aluminum	Per Floor	200.00	920.00
Bronze	Per Floor	200.00	1,130.00
RECEIVING BOX - Aluminum	Each	125.00	880.00
Bronze	Each	125.00	1,300.00
LETTER BOXES			
3" x 5" Alum. w/ Cam Locks Front Load	Each	10.00	60.00
Rear Load	Each	10.00	72.00
1022.0 SHOWER & TUB RECEPTORS & DOORS (M) (Cement Finishers)			
PRECAST TERRAZZO W/ STAINLESS STEEL CAP			
32" x 32" x 6"	Each	95.00	290.00
36" x 36" x 6"	Each	105.00	360.00
36" x 24" x 12"	Each	110.00	520.00
1023.0 SIGNS, LETTERS & PLAQUES (M) (Carpenters)			
LETTERS			
Cast Aluminum, Baked Enamel, Anodized Finish			
6" x 1/2" x 2"	Each	11.00	26.00
8" x 1/2" x 2"	Each	12.00	32.00
12" x 1/2" x 2"	Each	13.00	60.00
Plastic - Character	Each	2.50	3.50
PLAQUES - 24" x 24" Cast Aluminum	Each	100.00	860.00
24" x 24" Bronze	Each	100.00	1,000.00
SIGNS - HANDICAP (Metal)	Each	13.00	35.00
1024.0 STORAGE SHELVING (L&M) (Sheet Metal Workers)			
METAL - 18 Ga - 36" Sections 7'1" High			
12" Deep, 6 Shelves	LnFt	13.50	36.00
18" Deep, 6 Shelves	LnFt	14.50	43.00
24" Deep, 6 Shelves	LnFt	16.60	50.00
Add or Deduct per Shelf	LnFt	2.80	5.00
Add for Closed Back and Ends	LnFt	4.50	5.00
Add for 48' Sections	LnFt	4.60	3.50
Deduct for 22 Ga	LnFt	.50	3.50
PALLET RACKS - 6' x 36" - 4 Shelves	LnFt	28.00	100.00
1025.0 SUN CONTROL & PROTECTIVE COVERS (L&M)	SqFt	5.00	17.00
1026.0 TELEPHONE AND SOUND BOOTHS (M) (Carpenters)	Each	200.00	3,200.00
1027.0 TOILET & BATH ACCESSORIES (L&M) (Plumbers)			
BASED ON STAINLESS STEEL	UNIT	LABOR	MATERIAL
Curtain Rods - 60"	Each	18.00	30.00
Medicine Cabinets - 14" x 24"	Each	30.00	80.00
18" x 24"	Each	32.00	105.00
Mirrors - 16" x 10"	Each	20.00	65.00
18" x 24"	Each	22.00	80.00
24" x 30"	Each	30.00	100.00
Add for Shelves	-	-	27.00
Mop Holders	Each	25.00	70.00
Paper Dispenser - Surface Mounted	Each	20.00	30.00
Recessed	Each	25.00	60.00
Toilet Seat Covers - Recessed	Each	25.00	70.00

1027.0	TOILET & BATH ACCESSORIES, Cont'd...	UNIT	LABOR	MATERIAL
	Purse Shelf	Each	22.00	60.00
	Robe Hooks	Each	16.00	14.00
	Handicap Grab Bar - 18"	Each	21.00	35.00
	36"	Each	24.00	40.00
	42"	Each	30.00	45.00
	Set of Three	Each	65.00	90.00
	Soap Dispenser - Wall Mount	Each	20.00	45.00
	- with Lather Shelf	Each	22.00	90.00
	Soap and Grab Bar	Each	20.00	50.00
	Toilet Seat Cover Dispenser	Each	24.00	70.00
	Towel Bars - 24"	Each	20.00	30.00
	Towel Dispenser - Recessed	Each	45.00	400.00
	- Surface Mounted	Each	40.00	130.00
	and Disposal - Recessed	Each	38.00	400.00
	Sanitary Napkin Disposal - Recessed	Each	23.00	60.00
	Electric Hand Dryer	Each	60.00	330.00
1028.0	**TOILET COMPARTMENTS (Incl. Shower & Dressing Rooms) (L&M) (Sheet Metal Workers & Plumbers)**			
	PARTITIONS - FLOOR BRACED			
	(Door & 1 Side Panel) Baked Enameled	Each	120.00	380.00
	Stainless Steel	Each	125.00	900.00
	Plastic Laminate	Each	125.00	440.00
	Marble	Each	200.00	725.00
	Add for Extra Side Panel - Baked Enameled	Each	70.00	190.00
	Plastic Laminate	Each	70.00	210.00
	Marble	Each	140.00	270.00
	Add for Ceiling Hung	Each	70.00	100.00
	Add for Reinforcing for Handicap Bars (ADA)	Each	-	75.00
	Add for Paper Holders	Each	15.00	35.00
	Urinal Screens - Baked Enameled	Each	85.00	190.00
	Plastic Laminate	Each	85.00	210.00
	Marble	Each	125.00	340.00
	Shower Partitions (No Plumbing)			
	Baked Enameled (w/ Curtain & Terrazzo Base)			
	32" x 32"	Each	150.00	600.00
	36" x 36"	Each	150.00	540.00
	Fiberglass & Fiberglass Base - 32" x 32"	Each	90.00	240.00
	36" x 36"	Each	90.00	280.00
	Add for Doors	Each	40.00	190.00
	Tub Enclosures			
	Aluminum Frame and Tempered Glass Door	Each	70.00	170.00
	Chrome Frame and Tempered Glass Door	Each	70.00	230.00
1029.0	**URNS AND TRAYS**			
	CIGARETTE URNS - SS Rectangular or Round	Each	40.00	90.00
1030.0	**WALL AND CORNER GUARDS**			
	CORNER GUARDS			
	Stainless Steel			
	16 Ga 4" x 4" with Anchor	LnFt	6.00	19.00
	18 Ga 4" x 4" with Anchor	LnFt	5.50	18.50
	22 Ga 3½" x 3½" with Anchor	LnFt	6.00	15.50
	22 Ga 3½" x 3½" with Tape	LnFt	5.50	14.50
	Aluminum - Anodic with Tape	LnFt	5.50	13.00
	Vinyl	LnFt	5.00	8.25
	See 0502.3 for Steel			

		PAGE
1101.0	**MEDICAL CASEWORK (CSI 12342)**	**11-3**
.1	METAL	11-3
.2	PLASTIC - LAMINATE	11-3
.3	WOOD	11-3
.4	STAINLESS STEEL	11-3
1102.0	**EDUCATIONAL CASEWORK (CSI 12341)**	**11-3**
.1	METAL	11-3
.2	PLASTIC LAMINATE	11-3
.3	WOOD	11-3
1103.0	**KITCHEN AND BATH CASEWORK (CSI 11455)**	**11-3**
.1	PREFINISHED WOOD	11-3
.2	PLASTIC LAMINATE	11-3
1104.0	**MEDICAL EQUIPMENT (Hospital, Lab, Pharmacy) (CSI 11700)**	**11-3**
.1	STATIONS or SERVICE CENTERS	11-3
.2	LABORATORY TABLES and DEMONSTRATION DESKS (CSI 11600)	11-3
.3	TOPS and BOWLS (Special)	11-4
.4	FUME HOODS	11-4
.5	KEY CABINETS	11-4
.6	BLANKET and SOLUTION WARMERS	11-4
.7	ENVIRONMENTAL GROWTH CHAMBERS	11-4
.8	CHART RACKS and CHART RACK DESKS	11-4
.9	STERILIZING EQUIPMENT	11-4
.10	X-RAY and DARKROOM EQUIPMENT (CSI 11180)	11-5
.11	AUTOPSY and MORTUARY EQUIPMENT (CSI 11800)	11-5
.12	NARCOTIC SAFES	11-5
.13	REFRIGERATORS and FREEZERS (Other than Mortuary)	11-5
.14	TOTE TRAYS and UTILITY CARTS	11-5
.15	GAS TRACKS	11-5
.16	ANIMAL FENCING and CAGING	11-5
.17	DENTAL	11-5
.18	SURGICAL	11-5
.19	INCUBATORS	11-5
.20	PHYSICAL THERAPY	11-5
.21	OVERHEAD HOISTS	11-5
1105.0	**EDUCATIONAL EQUIPMENT (CSI 11300)**	**11-5**
1106.0	**RESIDENTIAL APPLIANCES (CSI 11452)**	**11-5**
.1	FREE STANDING	11-5
.2	BUILT IN	11-5
1107.0	**ATHLETIC, RECREATION & THERAPEUTIC EQUIPMENT (CSI 11480)**	**11-6**
.1	BASKETBALL BACKSTOPS	11-6
.2	GYM SEATING (Folding)	11-6
.3	SCOREBOARDS	11-6
.4	BOWLING and BILLIARDS	11-6
.5	SWIMMING POOL EQUIPMENT	11-6
.6	OUTSIDE RECREATIONAL EQUIPMENT	11-6
.7	GYMNASTIC EQUIPMENT	11-6
.8	SHOOTING RANGE EQUIPMENT (CSI 11496)	11-6
.9	HEALTH CLUB EQUIPMENT	11-6
1108.0	**FOOD SERVICE EQUIPMENT (CSI 11400)**	**11-7**
1109.0	**LAUNDRY AND DRY-CLEANING EQUIPMENT (CSI 11110)**	**11-7**
1110.0	**LIBRARY EQUIPMENT (CSI 11050)**	**11-7**

		PAGE
1111.0	**ECCLESIASTICAL EQUIPMENT (CSI 11040)**	**11-7**
.1	CROSSES, SPIRES and STEEPLES	11-7
.2	CONFESSIONALS	11-7
.3	PEWS	11-7
.4	BAPTISMAL FONTS	11-7
.5	ALTARS, LECTERNS, TABLES, etc.	11-7
1112.0	**STAGE AND THEATRICAL EQUIPMENT (CSI 11060)**	**11-8**
.1	STAGE DRAPERIES, HARDWARE, CONTROLS & OTHER EQUIPMENT	11-8
.2	FOLDING STAGES	11-8
.3	REVOLVING STAGES	11-8
.4	FIXED AUDITORIUM SEATING	11-8
.5	PROJECTION SCREENS	11-8
1113.0	**BANK EQUIPMENT (CSI 11030)**	**11-8**
.1	SAFES	11-8
.2	DRIVE-IN WINDOW UNITS	11-8
.3	TELLER UNITS	11-8
.4	NIGHT DEPOSITORIES	11-8
.5	SURVEILLANCE EQUIPMENT	11-8
.6	SAFE DEPOSIT BOX	11-8
1114.0	**DETENTION EQUIPMENT (CSI 11190)**	**11-8**
1115.0	**PARKING EQUIPMENT (CSI 11150)**	**11-8**
.1	AUTOMATIC DEVICES	11-8
.2	BOOTHS	11-8
.3	METERS and POSTS	11-8
1116.0	**LOADING DOCK EQUIPMENT (CSI 11160)**	**11-8**
.1	DOCK BUMPERS	11-8
.2	DOCK BOARDS	11-9
.3	LEVELERS and ADJUSTABLE RAMPS	11-9
.4	SHELTERS	11-9
1117.0	**SERVICE STATION EQUIPMENT**	**11-9**
.1	REELS	11-9
.2	PUMPS	11-9
.3	COMPRESSORS	11-9
.4	TIRE CHANGERS	
1118.0	**MUSICAL EQUIPMENT (CSI 11070)**	**11-9**
1119.0	**CHECKROOM EQUIPMENT (CSI 11090)**	**11-9**
1120.0	**AUDIO VISUAL EQUIPMENT (CSI 11130)**	**11-9**
1121.0	**INCINERATORS (CSI 11171)**	**11-9**
1122.0	**PHOTO AND GRAPHIC ART EQUIPMENT (CSI 11470)**	**11-9**
1123.0	**WASTE HANDLING EQUIPMENT (11860)**	**11-9**

1101.0 MEDICAL CASEWORK (L&M) (Carpenters)
INCLUDING PLASTIC LAMINATE TOP, HARDWARE & FINISH

		UNIT	LABOR	MATERIAL
.1	METAL			
	Base Cabinets - 35" High x 24" Deep	LnFt	29.00	180.00
	Wall Cabinets - 24" High x 12" Deep	LnFt	27.00	140.00
	High Wall (Utility) - 84" High x 12" Deep	LnFt	30.00	145.00
.2	PLASTIC LAMINATE (Including Plastic Laminator)			
	Base Cabinets - 35" High x 24" Deep	LnFt	29.00	190.00
	Wall Cabinets - 24" High x 12" Deep	LnFt	27.00	130.00
	High Wall (Utility) - 84" High x 12" Deep	LnFt	30.00	155.00
.3	WOOD			
	Base Cabinets - 35" High x 24" Deep	LnFt	29.00	175.00
	Wall Cabinets - 24" High x 12" Deep	LnFt	27.00	130.00
	High Wall (Utility) - 84" High x 12" Deep	LnFt	31.00	150.00
.4	STAINLESS STEEL			
	Base - 35" High x 24" Deep	LnFt	30.00	210.00
	Wall - 24" High x 12" Deep	LnFt	27.00	195.00

1102.0 EDUCATIONAL CASEWORK (L&M) (Carpenters)
Same as 1101.0 Medical Casework above.

1103.0 KITCHEN AND BATH CASEWORK (M) (Carpenters)

		UNIT	LABOR	MATERIAL
.1	PREFINISHED WOOD - NO TOPS			
	Base Cabinets - 35" High x 18" Deep	LnFt	29.00	150.00
	Front Only	LnFt	27.00	85.00
	with 1 Drawer	LnFt	29.00	180.00
	with 4 Drawers	LnFt	30.00	240.00
	with Lazy Susan	LnFt	29.00	190.00
	Upper Cabinets - 30" High x 12" Deep	LnFt	28.00	120.00
	18" High x 12" Deep	LnFt	27.00	95.00
	with Lazy Susan	LnFt	28.00	160.00
	High Cabinets - 84" High x 18" Deep	LnFt	30.00	240.00

Add for Counter Tops - See 0603.2

		UNIT	LABOR	MATERIAL
.2	PLASTIC LAMINATE			Add 10%

1104.0 MEDICAL EQUIPMENT (M) (Carpenters)

		UNIT	LABOR	MATERIAL
.1	STATION AND SERVICE CENTER EQUIPMENT			
.11	Kitchen -			
	Example: 4'6" x 4'2" Base & 30" Upper	Each	210.00	1,850.00
.12	Medical Preparation -			
	Example: 4'0" x 6'8" x 1'8"	Each	260.00	4,400.00
	(also 5' and 6' wide)			
.13	Nourishment and Ice Stations -			
	Example: 6'0" x 6'8" x 2'8"	Each	275.00	9,400.00
.14	Lavatory Unit -			
	Example: 4'7" x 1'8" x 4'0"	Each	130.00	875.00
.15	Janitor's Closet -			
	Example: 2'0" x 5'4" x 6'8"	Each	240.00	3,700.00
.16	Glove Packaging - Example	Each	130.00	960.00
.17	Linen Inspection - Example	Each	150.00	1,400.00
.2	LAB TABLES AND DEMONSTRATION DESKS	Each	140.00	2,100.00

1104.0 MEDICAL EQUIPMENT, Cont'd...

		UNIT	LABOR	MATERIAL
.3	**TOPS AND BOWLS (M) (Carpenter or Plumber)**			
	Calcium Aluminum Silicate	SqFt	5.50	22.00
	Corian	SqFt	5.50	42.00
	Epoxy Resins (Kemstone)	SqFt	5.50	20.00
	Marble - Artificial	SqFt	4.40	22.00
	Plastic Laminate - 25" - 1 1/2"	SqFt	3.75	20.00
	Add for Backsplash	LnFt	1.60	4.00
	Soapstone	SqFt	4.30	30.00
	Stainless Steel - Bowl	Each	63.00	170.00
	Tops	SqFt	5.70	55.00
	Add for Backsplash	LnFt	1.80	7.00
	Vinyl Sheet - 25" - 1 1/2"	SqFt	2.80	12.00
	Add for Backsplash	LnFt	1.50	4.00
.4	**FUME HOODS (INCLUDING DUCT WORK)**			
	Stock - 4"	Each	380.00	4,600.00
	Custom	Each	380.00	5,100.00
.5	**KEY CABINETS (100 KEY)**	Each	75.00	270.00
.6	**BLANKET AND SOLUTION WARMERS**			
	(NO MECHANICAL OR ELECTRICAL)			
	Example: 27 x 11 x 30" x 74			
	Free Standing - Steam	Each	350.00	4,400.00
	Electric	Each	350.00	4,600.00
	Recessed - Steam	Each	400.00	4,300.00
	Electric	Each	400.00	4,400.00
.7	**ENVIRONMENTAL GROWTH CHAMBERS & LABS**	Each	1,000.00	4,100.00
.8	**CHART RACKS AND CHART RACK DESKS**			
	Racks	Each	65.00	270.00
	Desks	Each	100.00	1,350.00

			ELECTRIC		STEAM	
.9	**STERILIZING EQUIPMENT**		Re-cessed	Open Mtd or Mobile	Re-cessed	Open Mtd or Mobile
	(NO MECHANICAL/ ELECTRICAL)	UNIT				
	Dressing and Instrument -					
	Example: 16" x 26"	Each	9,400	-	850	-
	Dressing -					
	Example: 20" x 20" x 38"	Each	19,000	-	15,000	1,550
	Solution Warming/ Storage Cab.	Each	-	-	-	-
	Bedpan -					
	Example: 16" x 16" x 27"					
	Non Pressure	Each	-	2,200	-	-
	Pressure	Each	-	2,500	-	-
	Laboratory (Painted)	Each	10,800	11,400	9,700	11,800
	Add for Stainless Steel					
	Autoclave (All Purpose)					
	Portable Inst. - Non Pressure					
	6" x 6" x 13"	Each	-	925	-	-
	8" x 8" x 16"	Each	-	1,480	-	-
	10" x 10" x 22"	Each	-	4,700	-	-
	Hot Air -					
	Example: 39" x 14" x 19"					
	Gravity	Each	2,700	2,300	-	-
	Convection	Each	4,000	4,200	-	-

1104.0 MEDICAL EQUIPMENT, Cont'd...

.10 X-RAY AND DARK ROOM EQUIPMENT

Lead Protection - See 0801 for Doors & View Window Frames

Lead Protection - See 0902 for Lead Lined Sheetrock

Manual Dark Room

	UNIT	LABOR	MATERIAL
Example: Room including Cassette Pass Box, Cassette Storage Cabinet, Film Identifier Cabinet, Film Illuminator and Dryer, Development Tank including Tops and Splashes	Each	1,400.00	14,000.00

		UNIT	LABOR	MATERIAL
	Automatic			
	Processor with Solution Storage Tank	Each	750.00	10,500.00
	Film Loader	Each	250.00	5,200.00
.11	AUTOPSY AND MORTUARY EQUIPMENT			
	Refrigerators	Each	425.00	8,700.00
	Tables	Each	270.00	5,400.00
	Carts	Each	80.00	1,450.00
.12	NARCOTIC SAFES	Each	90.00	550.00
.13	REFRIGERATORS (OTHER THAN MORTUARY)			
	Under Counter	Each	210.00	2,300.00
.14	TOTE TRAYS AND UTILITY CARTS	Each	75.00	450.00
.15	GAS TRACKS - Example: Operating Room	Each	440.00	3,000.00
	Straight	LnFt	40.00	240.00
	Curved	LnFt	50.00	295.00
.16	ANIMAL FENCING AND CAGING	SqFt	2.50	5.60
.17	DENTAL	-	-	-
.18	SURGICAL	-	-	-
.19	INCUBATORS	Each	-	4,200.00
.20	THERAPY-HEAT	Each	-	2,050.00
.21	OVERHEAD HOISTS	Each	-	2,000.00

1105.0 EDUCATIONAL EQUIPMENT

1106.0 RESIDENTIAL EQUIPMENT (NO MECH. OR ELEC.)

See 1504 for Supply, Waste and Vent

See 1611 for Electric Power

.1 FREE STANDING

		UNIT	LABOR	MATERIAL
	Clothes Dryer - Gas	Each	80.00	440.00
	Electric	Each	80.00	390.00
	Clothes Washers	Each	80.00	410.00
	Compactors - 15"	Each	75.00	470.00
	Disposals	Each	75.00	155.00
	Dishwasher	Each	80.00	460.00
	Freezers - Upright - 16 CuFt	Each	80.00	410.00
	19 CuFt	Each	85.00	460.00
	21 CuFt	Each	95.00	490.00
	Freezers - Upright - Frost Proof			
	15 CuFt	Each	80.00	510.00
	18 CuFt	Each	80.00	560.00
	Freezers - Chest - 15 CuFt	Each	80.00	460.00
	20 CuFt	Each	90.00	560.00
	25 CuFt	Each	100.00	650.00
	Dehumidifier - 40 Pint	Each	80.00	170.00
	Humidifier - 10 Gals/Day	Each	80.00	230.00
	Range Hood	Each	70.00	165.00
	Water Heater - 30 Gal. & 40 Gal.	Each	120.00	215.00
	Water Softener - 30 Grains	Each	150.00	630.00

1106.0 RESIDENTIAL EQUIPMENT, Cont'd...

	UNIT	LABOR	MATERIAL
.1 FREE STANDING, Cont'd.			
Refrigerators			
Compact - 11 CuFt	Each	80.00	330.00
Conventional- 14 CuFt	Each	85.00	450.00
With Freezer- 15 CuFt (Also 17 & 19)	Each	90.00	550.00
21 CuFt	Each	95.00	700.00
Frostproof- 12 CuFt	Each	85.00	570.00
15 CuFt	Each	95.00	760.00
Side-by-Side- 22 CuFt	Each	95.00	1,050.00
25 CuFt	Each	100.00	1,200.00
Ranges - Electric - 30"	Each	80.00	620.00
40"	Each	85.00	1,250.00
Gas - 20"	Each	75.00	270.00
30"	Each	80.00	630.00
Microwave Ovens - Counter	Each	80.00	180.00
Over-counter	Each	120.00	420.00
.2 BUILT-IN			
Air Conditioners - 110 Volt	Each	75.00	350.00
220 Volt	Each	75.00	530.00
Dishwashers	Each	75.00	450.00
Disposals	Each	75.00	159.00
Hoods	Each	75.00	170.00
Ranges - Surface - 30" Gas	Each	75.00	270.00
42" Gas with Broiler	Each	75.00	1,230.00
30" Electric	Each	75.00	280.00
42" Elec. with Broiler	Each	80.00	1,230.00
Ovens - Single - Gas	Each	80.00	420.00
Double	Each	90.00	530.00
Single - Electric	Each	90.00	380.00
Double	Each	90.00	480.00

1107.0 ATHLETIC EQUIPMENT (M) (Carpenters)

	UNIT	LABOR	MATERIAL
.1 BASKETBALL BACKSTOPS			
Stationary	Each	350.00	640.00
Retractable	Each	500.00	1,550.00
Suspended	Each	600.00	3,000.00
Add for Electric Operated	Each	-	925.00
.2 GYM SEATING - See Div. 1306.0			
.3 SCOREBOARDS (Great Variable)	Each	950.00	5,600.00
.4 BOWLING AND BILLIARDS - See Div. 1302.0			
.5 SWIMMING POOL EQUIPMENT- See Div. 1318.0			
.6 OUTSIDE RECREATIONAL EQ.- See Div. 0208.2			
.7 GYMNASTIC EQUIPMENT - See Div. 0208.2			
.8 SHOOTING RANGE EQUIPMENT	Point	700.00	6,000.00
.9 HEALTH CLUB EQUIPMENT	-	-	-

1108.0 FOOD SERVICE EQUIPMENT (L&M) (Sheet Metal and Plumbing)

	UNIT		COST
Broilers - 36" Electric	Each	-	4,500.00
Can Washers	Each	-	3,250.00
Coffee Stand and Urn	Each	-	4,000.00
Cold Food Carts	Each	-	4,300.00
Conveyors - 12'	Each	-	5,700.00
Cutter - Mixer	Each	-	6,750.00
Dishtables - Clean	Each	-	3,500.00
Soiled	Each	-	4,700.00

1108.0 FOOD SERVICE EQUIPMENT, Cont'd...

	UNIT	COST
Disposer - Garbage - 3 H.P.	Each	3,700.00
1 1/2 H.P.	Each	2,100.00
Dishwasher- Automatic	Each	17,500.00
Rack Type	Each	6,200.00
Fryers - Double	Each	2,200.00
Hot Food Carts	Each	3,400.00
Ice Cream Dispenser	Each	1,250.00
Ice Machine with Bin	Each	1,800.00
Juice and Beverage Dispenser	Each	1,900.00
Meat Saw	Each	3,200.00
Microwave	Each	310.00
Oven - Double Convector - Electric	Each	6,500.00
Single	Each	3,300.00
Portable Sink	Each	840.00
Range - Oven- Four Burner	Each	1,350.00
Range Hoods	Each	13,600.00
Refrigerators	Each	2,300.00
Silverware Dispensers	Each	215.00
Sinks - Utility	Each	1,150.00
Soiled Dish	Each	4,100.00
Slicer	Each	2,300.00
Steamers	Each	4,700.00
Tray Dispensers	Each	670.00
Setup Unit	Each	1,250.00
Utility Carts	Each	200.00
Water Stations	Each	1,025.00
Work Tables	Each	1,050.00
Average per Kitchen for Equipment	SqFt	60.00

1109.0 LAUNDRY EQUIPMENT (L&M)

	UNIT	COST
Dryer - 50 Lb.	Each	1,900.00
Ironer, Air	Each	6,700.00
Hand	Each	4,600.00
Washer Tumbler - 135 Lb.	Each	32,000.00
50 Lb.	Each	17,500.00

1110.0 LIBRARY EQUIPMENT (M) (Carpenters)

		LABOR	MATERIAL
Book Trucks	Each	90.00	1,050.00
Card Catalog Index - 30 drawer	Each	115.00	1,850.00
Chairs	Each	15.00	100.00
Charging Desks - Card File	Each	90.00	940.00
Charging Unit	Each	90.00	1,000.00
Open Shelf	Each	90.00	670.00
Newspaper Rack	Each	85.00	650.00
Shelving (Maple) - Example: 36" x 60" x 8"	LnFt	15.00	100.00
Tables	Each	65.00	665.00

1111.0 ECCLESIASTICAL EQUIPMENT (M) (Carpenters)

			LABOR	MATERIAL
.1	CROSSES, SPIRES AND STEEPLES	-	-	-
.2	CONFESSIONALS - Two Person	Each	540.00	5,000.00
	One Person	Each	425.00	3,500.00
.3	PEWS (OAK) (With Kneelers)	LnFt	13.70	100.00
	Add for Frontals	LnFt	13.80	50.00
	Add for Padded Kneelers	LnFt	3.80	3.00
	Add for Birch	LnFt	-	10.00
.4	BAPTISMAL FONTS	Each	70.00	710.00
.5	MISCELLANEOUS (WOOD ONLY)			
	Main Altars	Each	420.00	5,000.00
	Side Altars	Each	235.00	1,900.00
	Lecterns	Each	140.00	1,050.00
	Credence Table	Each	110.00	740.00

1112.0 STAGE AND THEATRICAL EQUIPMENT (L&M) (Carpenters)

		UNIT	LABOR	MATERIAL
.1	STAGE DRAPERIES, HARDWARE, CONTROLS AND OTHER EQUIPMENT			
	Motorized Rolling Curtain - 38' x 16'	Each	-	5,400.00
	Stage Equipment Sets	Each	-	1,850.00
	Light Bridge	Each	-	19,500.00
.2	FOLDING STAGES - Portable Thrust	Each	-	12,300.00
.3	REVOLVING STAGES - 32' Dia x 12" High	Each	-	120,000.00
.4	FIXED AUDITORIUM SEATING - Cushioned	Each	-	170.00
.5	PROJECTION SCREENS			
	Motor Oper. - 8' x 8'	Each	120.00	1,400.00
	10' x 10'	Each	125.00	1,650.00
	Stationary - 20' x 20'	Each	150.00	2,700.00

1113.0 BANK EQUIPMENT (M) (Carpenters)

		UNIT	LABOR	MATERIAL
.1	SAFES	Each	140.00	1,000.00
.2	DRIVE-UP WINDOW UNIT - Including Cash Drawers, Counter, Window, Depository and Lock Box	Each	870.00	8,000.00
.3	TELLER UNIT	Each	290.00	2,200.00
	Cash Dispensing	Each	2,750.00	37,000.00
.4	NIGHT DEPOSITORY - Envelope	Each	190.00	970.00
	Incl. Head & Chest	Each	900.00	9,000.00
.5	SURVEILLANCE SYSTEMS			
	Receival and Transmitter	Each	-	650.00
	Cameras	Each	-	970.00
	Time Lapse Recorder	Each	-	3,200.00
.6	SAFE DEPOSIT BOX	Each	6.90	70.00
	Vault Doors - See Div. 0803.16			
	Pneumatic Tube System - See Div. 14			

1114.0 DETENTION EQUIPMENT

1115.0 PARKING EQUIPMENT

		UNIT	LABOR	MATERIAL
.1	AUTOMATIC DEVICES			
	Cash Registers	Each	510.00	10,500.00
	Clocks - Rate Computing	Each	82.00	775.00
	Standards	Each	67.00	620.00
	Gate	Each	380.00	3,000.00
	Loop Sensor	Each	200.00	925.00
	Treadle	Each	155.00	370.00
	Operator Station - Coin	Each	155.00	4,650.00
	Key	Each	170.00	1,000.00
	Radio Control - Transmitter	Each	-	1,700.00
	Ticket Spitter Station	Each	215.00	5,500.00
.2	BOOTHS - Example: 4' x 6'	Each	250.00	5,400.00
.3	METERS AND POSTS	Each	80.00	300.00

1116.0 LOADING DOCK EQUIPMENT

		UNIT	LABOR	MATERIAL
.1	DOCK BUMPERS			
	16" x 12" x 4"	Each	22.00	48.00
	24" x 6" x 4"	Each	22.00	49.00
	24" x 10" x 4"	Each	23.00	53.00
	24" x 12" x 4"	Each	23.00	71.00
	36" x 6" x 4"	Each	24.00	64.00
	36" x 10" x 4"	Each	26.00	79.00
	36" x 12" x 4"	Each	30.00	92.00
	Add for Galvanized Angles	Each	-	6.25

		UNIT	LABOR	MATERIAL
1116.0	**LOADING DOCK EQUIPMENT, Cont'd...**			
.2	DOCK BOARDS -			
	5,000# 3' - 6" x 6' - 0"	Each	110.00	890.00
	10,000# 4' - 0" x 6' - 0"	Each	120.00	1,140.00
	15,000# 4' - 0" x 6' - 0"	Each	130.00	1,500.00
.3	LEVELERS & ADJUSTABLE RAMPS			
	8' - 0" x 8' (Add Concrete)	Each	215.00	2,700.00
.4	SHELTERS - 10' x 10' with Dock Pad	Each	290.00	1,700.00
	Strip Curtain - 10' x 10' x 8"	Each	95.00	220.00
	Dock Pad - 10' x 10'	Each	175.00	600.00
1117.0	**SERVICE STATION EQUIPMENT**			
.1	REELS - Chassis (No Mech. or Electric)	Each	120.00	600.00
	Air	Each	120.00	500.00
	Water	Each	120.00	520.00
	Motor Oil	Each	120.00	610.00
	A.T.F.	Each	120.00	660.00
	Gear Lube	Each	120.00	600.00
	Combination of Above - 3 Hose	Each	500.00	4,900.00
	Combination of Above - 5 Hose	Each	800.00	6,000.00
.2	PUMPS - Chassis	Each	140.00	530.00
	Gear Oil	Each	140.00	590.00
	A.T.F.	Each	140.00	590.00
	Oil - Motor	Each	100.00	630.00
	Gasoline	Each	300.00	1,850.00
.3	COMPRESSORS - 5 H.P.	Each	300.00	4,100.00
	3 H.P.	Each	290.00	1,950.00
	1 1/2 H.P.	Each	275.00	1,830.00
.4	TIRE CHANGERS - AUTO	Each	390.00	2,750.00
	Exhaust Systems - See Div. 15			
	Hoisting Equipment - See Div. 14			
	Tanks/Air/Water Connections - See Div. 15			
1118.0	**MUSICAL EQUIPMENT**			
1119.0	**CHECKROOM EQUIPMENT**			
	MOTORIZED, 1-Shelf, 16 ft	LnFt	18.00	110.00
	2-Shelf, 16 ft	LnFt	25.00	220.00
1120.0	**AUDIO VISUAL EQUIPMENT**			
	PROJECTION SCREEN	Each	-	120.00
	ELECTRIC CONTROLLED	Each	-	480.00
	TELEVISION CAMERA	Each	-	930.00
	MONITOR	Each	-	680.00
	VIDEO TAPE RECORDER	Each	-	980.00
1121.0	**INCINERATORS**			
	SMALL - 18" x 18"	Each	220.00	1,950.00
	MEDIUM - 22" x 22"	Each	240.00	2,800.00
	LARGE	Each	340.00	4,000.00
1122.0	**PHOTO AND GRAPHIC ART EQUIPMENT**			
1123.0	**WASTE HANDLING EQUIPMENT**			
	COMPACTORS - Bag - 3 CuYd Hopper	Each	850.00	10,700.00
	Cart	Each	850.00	9,500.00

		PAGE
1201.0	**MANUFACTURED CASEWORK (CSI 12300)**	**12-1**
	(Including Plastic Laminate Top and Hardware)	
.1	MEDICAL	12-1
.11	Wood - Prefinished	12-1
.12	Plastic Laminate	12-1
.13	Enameled Metal	12-1
.14	Stainless Steel	12-1
.2	EDUCATIONAL	12-1
1202.0	**BLINDS, SHADES AND SHUTTERS (CSI 12505)**	**12-2**
.1	BLINDS	12-2
.2	SHADES	12-2
.3	SHUTTERS	12-2
1203.0	**DRAPERIES AND CURTAIN HARDWARE (CSI 12530)**	**12-2**
1204.0	**OPEN OFFICE FURNITURE (CSI 12610)**	**12-2**
1205.0	**FLOOR MATS AND FRAMES (CSI 12690)**	**12-2**
1206.0	**AUDITORIUM AND THEATRE SEATING (CSI 12710)**	**12-2**
1207.0	**MULTIPLE USE SEATING (CSI 12710)**	**12-2**
1208.0	**BUILT-IN FOLDING TABLES AND SEATING (CSI 12745)**	**12-2**
.1	TABLES	12-2
.2	CHAIRS	12-2

1201.0 MANUFACTURED CASEWORK (L&M) (Carpenters)
(Including Plastic Laminate Top & Hardware)

		UNIT	LABOR	MATERIAL
.1	MEDICAL CASEWORK			
.11	Metal			
	Base Cabinets - 35" High x 24" Deep	LnFt	31.00	185.00
	Upper Cabinets - 24" High x 12" Deep	LnFt	30.00	150.00
	High Wall - 84" High x 12" Deep	LnFt	31.00	155.00
.12	Plastic Laminate			
	Base Cabinets - 35" High x 24" Deep	LnFt	31.00	200.00
	Upper Cabinets - 24" High x 12" Deep	LnFt	30.00	145.00
	High Wall - 84" High x 12" Deep	LnFt	31.00	230.00
.13	Wood - Prefinished			
	Base Cabinets - 35" High x 24" Deep	LnFt	31.00	185.00
	Upper Cabinets - 24" High x 12" Deep	LnFt	30.00	140.00
	High Wall - 84" High x 12" Deep	LnFt	30.00	225.00
.14	Stainless Steel			
	Base - 35" High x 24" Deep	LnFt	32.00	230.00
	Wall - 24" High x 12" Deep	LnFt	30.00	210.00
.2	EDUCATIONAL CASEWORK			

DIVISION #12 - FURNISHINGS

1202.0 BLINDS, SHADES AND SHUTTERS (L&M) (Carpenters)

	UNIT	COST
.1 BLINDS - Horizontal		
Wood 1"	SqFt	5.40
Aluminum 1"	SqFt	5.20
1"	SqFt	4.15
Steel 2"	SqFt	3.75
Cloth 1"	SqFt	6.25
Vertical		
Cloth or PVC 3"	SqFt	5.20
Aluminum 3"	SqFt	6.50
.2 SHADES - Cotton	SqFt	1.85
Vinyl Coated	SqFt	2.10
Lightproof	SqFt	2.70
Slat	SqFt	2.30
Woven Aluminum	SqFt	4.00
Fibre Glass	SqFt	2.80
X-ray and Dark Room	SqFt	9.80
.3 SHUTTERS		
16" x 1 1/8" x 48"	Pair	98.00
16" x 1 1/8" x 60"	Pair	115.00
16" x 1 1/8" x 72"	Pair	120.00

1203.0 DRAPERIES AND CURTAIN HARDWARE (L&M) (Carpenters)

	UNIT	COST
TRACK	LnFt	4.60
DRAPERIES	SqYd	8.25
Add for Lined	SqYd	1.85
Add for Rods	SqYd	2.10
Add for Motorized	SqYd	6.20

1204.0 OPEN OFFICE FURNITURE

1205.0 FLOOR MATS AND FRAMES (M) (Carpenters)

	UNIT	LABOR	MATERIAL
MATS - Vinyl Link - 4' x 5' x 1/2"	Each	30.00	310.00
4' x 7' x 1/2"	Each	30.00	430.00
Add for Color	SqFt	-	4.00
FRAMES - Aluminum - 4' x 5'	Each	60.00	195.00
4' x 7'	Each	70.00	240.00

1206.0 AUDITORIUM AND THEATRE SEATING (L&M) (Carpenters)

	UNIT	COST
WOOD AND PLYWOOD	Each	135.00
CUSHIONED	Each	165.00
ROCKING	Each	245.00

1207.0 MULTIPLE-USE SEATING

	UNIT	COST
MOVABLE - Chrome and Solid Plastic	Each	65.00
Wood and Plywood	Each	58.00

1208.0 BUILT-IN FOLDING TABLES AND SEATING (M) (Carpenters)

	UNIT	COST
.1 TABLES - 3' x 8' - Tempered Hardboard Top	Each	180.00
Vinyl Laminate Top	Each	205.00
Folding Table & Bench-Pocket Unit	Each	1,530.00
.2 CHAIRS - Metal	Each	20.00

		PAGE
1301.0	**AIR SUPPORTED STRUCTURE (CSI 13010)**	**13-2**
.1	TENNIS COURTS	13-2
.2	CONSTRUCTION DOMES, WAREHOUSES, etc.	13-2
1302.0	**BOWLING ALLEYS (CSI 11500)**	**13-2**
1303.0	**CHIMNEYS - SPECIAL**	**13-2**
.1	JOB CONSTRUCTED (Radial Brick and Reinforcing Concrete (CSI 04550)	13-2 13-2
.2	PREFABRICATED (Class A Metal Refractory Lined) (CSI 15600)	13-2
1304.0	**CLEAN ROOMS (CSI 13250)**	**13-2**
1305.0	**FLOORS - PEDESTAL OR ACCESS (CSI 10270)**	**13-2**
1306.0	**GRANDSTANDS, BLEACHERS AND GYM SEATING (CSI 13125)**	**13-3**
.1	GRANDSTANDS and BLEACHERS	13-3
.2	GYM SEATING - FOLDING	13-3
1307.0	**GREENHOUSES (CSI 13123)**	**13-3**
1308.0	**INCINERATORS (CSI 13400)**	**13-3**
1309.0	**COLD STORAGE ROOMS (CSI 13038)**	**13-3**
.1	CUSTOM - JOB CONSTRUCTED	13-3
.2	PREFABRICATED	13-3
1310.0	**METAL BUILDINGS (CSI 13122)**	**13-4**
1311.0	**RADIATION PROTECTION & RADIO FREQUENCY SHIELDING (CSI 13090 & 13095)**	**13-4**
.1	ELECTROSTATIC	13-4
.2	ELECTRO-MAGNETIC (Copper and Steel - Solid and Screen)	13-4
.3	X-RAY (Nuclear and Gamma)	13-4
.4	RADIO FREQUENCY SHIELDING	13-4
1312.0	**SAUNAS (CSI 13052)**	**13-4**
1313.0	**SCALES (CSI 10650)**	**13-5**
.1	FLOOR SCALES	13-5
.2	TRUCK SCALES	13-5
.3	CRANE SCALES	13-5
1314.0	**SIGNS (Custom Exterior)**	**13-5**
1315.0	**DOME STRUCTURES (Including Observatories) (CSI 13132)**	**13-5**
1316.0	**SOUND INSULATED ROOMS (CSI 13034)**	**13-5**
.1	ANECHOIC ROOMS	13-5
.2	AUDIOMETRIC ROOMS	13-5
1317.0	**SOUND AND VIBRATION CONTROL SYSTEMS (CSI 13081)**	**13-5**
1318.0	**SWIMMING POOLS (CSI 13152)**	**13-6**
.1	POOLS	13-6
.2	DECKS	13-6
.3	FENCES	13-6
1319.0	**TANKS**	**13-6**
.1	STEEL - GROUND LEVEL	13-6
.2	STEEL - ELEVATED	13-6
1320.0	**WOOD DECKS**	**13-6**

1301.0 AIR SUPPORTED STRUCTURES

		UNIT	COST
.1	TENNIS COURTS - 2 LAYER THERMAL DOMES (Includes Doors, Heating, Equipment, Lights & Power)		
	1 Court - 58' x 118'	SqFt	17.00
	2 Courts - 106' x 118'	SqFt	16.00
	3 Courts - 156' x 118'	SqFt	14.50
	4 Courts - 201' x 118'	SqFt	13.50
	Deduct for Non Thermal Type	SqFt	2.00
.2	CONSTRUCTION DOMES, WAREHOUSES, ETC.	SqFt	7.50
	Add for Heating	SqFt	3.00
	Add for Light and Power	SqFt	1.50
	Add for Air Lock Doors	Each	13,300.00
	Add for Revolving Doors	Each	6,800.00

1302.0 BOWLING ALLEYS (L&M) (Carpenters)

	UNIT	COST
BOWLING ALLEYS (L&M) (Carpenters)	Lane	37,000.00
Add for Automatic Scorers	Lane	14,500.00

1303.0 CHIMNEYS - SPECIAL (Foundations Not Included)

.1 JOB CONSTRUCTED (L&M) (Bricklayers)

Inside Diam.	Height		UNIT	COST
5'	100'	Radial Brick	Each	120,000.00
6'	150'	Radial Brick	Each	145,000.00
7'	200'	Radial Brick	Each	190,000.00
8'	250'	Reinforced Concrete	Each	460,000.00
9'	200'	Reinforced Concrete	Each	570,000.00
10'	250'	Reinforced Concrete	Each	780,000.00
12'	300'	Reinforced Concrete	Each	1,300,000.00

.2 PREFABRICATED (METAL REFRACTORY LINED) 1800° - 2000° (L&M) (Sheet Metal & Iron Workers)

	UNIT	COST
10"	LnFt	54.00
12"	LnFt	58.00
15"	LnFt	69.00
18"	LnFt	84.00
21"	LnFt	100.00
24"	LnFt	120.00
30"	LnFt	215.00
36"	LnFt	300.00
Add for Cleanouts or T's	Each	310.00
Add for V.L. Label	Each	10%
Deduct for 800° Chimney	Each	5%

1304.0 CLEAN ROOMS

	UNIT	COST
CLEAN ROOMS	SqFt	180.00

1305.0 FLOORS - ACCESS OR PEDESTAL (L&M) (Carpenters)
(Includes Laminate Topping)

	UNIT	COST
Aluminum	SqFt	19.00
Steel - Galvanized	SqFt	12.00
Plywood - Metal Covered	SqFt	11.00
Add for Carpeting	SqFt	3.00
Add for Stairs and Ramps	SqFt	50%

1306.0 GRANDSTANDS, BLEACHERS AND GYM SEATING

		UNIT	COST
.1	GRANDSTANDS AND BLEACHERS		
	Not Permanent	Each Seat	23.50
	Permanent - Aluminum	Each Seat	94.00
	Steel	Each Seat	60.00
	Concrete	Each Seat	220.00
.2	GYM SEATING - FOLDING	Each Seat	66.00

1307.0 GREENHOUSE (L&M) (Glaziers)

	UNIT	COST
SINGLE ROOM TYPE - Commercial	SqFt	21.00
Educational	SqFt	28.00
Residential	SqFt	31.00
Add for Heating, Lighting and Benches	SqFt	17.00
Add for Cooling	SqFt	7.20

1308.0 INCINERATORS (L&M) (Ironworkers and Bricklayers)

Stack and Breaching Not Included
Pounds/Hour

	UNIT	COST
100	Each	8,100.00
200	Each	8,200.00
300	Each	9,100.00
400	Each	10,300.00
500	Each	11,300.00
600	Each	14,500.00
800	Each	23,500.00
Add for Gas Burner	Each	1,540.00
Add for Controls	Each	850.00
Add for Piping	Each	870.00

1309.0 INSULATED ROOMS (L&M) (Carpenters)

		UNIT	COST
.1	CUSTOM - Cooler	SqFt	165.00
	Freezer	SqFt	190.00
.2	PREFABRICATED (SELF-CONTAINED)		

	UNIT	COST
Cooler Example: Room 7'6" x 9'6"		
Wood (Plywood)	Each	6,000.00
Metal Covered 20 ga Galvanized #3 SS	Each	7,400.00
Add for 3/4 HP Compressor	Each	1,000.00
Add for 3/4 HP Compressor Starter & Coils	Each	4,550.00
Add for Electrical	Each	775.00
Cooler - Total Cost - Wood	SqFt	165.00
Metal Covered	SqFt	185.00
Freezer Example: Room 7'6" x 9'6"		
Wood (Plywood)	Each	6,150.00
Metal Covered 20 ga Galvanized #3 SS	Each	7,700.00
Add for 3/4 HP Compressor	Each	1,340.00
Add for 3/4 HP Compressor Starter & Coils	Each	4,150.00
Add for Electrical	Each	800.00
Freezer - Total Cost - Wood	SqFt	180.00
Metal Covered	SqFt	210.00
Add for Shelving	SqFt	8.50

		UNIT	COST
1310.0	**METAL BUILDINGS**		
.1	SHELL - FLOOR AREA (NO FOUNDATION)	SqFt	13.50
	24 ga. 12' Sidewalls and 60' Span (Galvanized)		
	Add per foot above 12' - Floor Area	SqFt	.38
	Add per foot of Span above or below 60'	SqFt	.23
	Add for 2' Overhand - Overhang Area	SqFt	5.20
	Add for Aluminum	SqFt	.65
	Add for Liner Panels - Wall Area	SqFt	1.95
	Add for Insulation 3 1/2" - Wall Area	SqFt	.75
	Add for Insulation 6" - Ceiling Area	SqFt	.95
	Add for Door Openings - Including Hardware	Each	670.00
	Add for Window Openings	Each	360.00
	Add for Insulated Glass Area	SqFt	10.00
	Add for Overhead Door Area	SqFt	9.25
	See 0803.4 for Door Cost		
	Add for Gutter	LnFt	4.50
	FOUNDATION & GROUND SLAB COST AVG (3-Foot Frost)	SqFt	7.80
1311.0	**RADIATION PROTECTION (L&M) (Carpenters)**		
.1	ELECTROSTATIC SHIELDING		
	Custom - Vinyl Sandwiched Copper Mesh for Floors, Walls & Ceilings (including Copper Foil Tabs & Plastic Tape)	SqFt	23.00
	Add per door and Frame with RF Shielded Glass (including Grounding Wires)	Each	3,500.00
	Add per Borrowed Light RF Shielded Glass	Each	1,400.00
	Example: Room 10' x 8' (Prefab)	Room	27,000.00
		or SqFt	350.00
.2	ELECTROMAGNETIC SHIELDING		
	Example: Room 10' x 8' (Prefab)	Room	11,500.00
		or SqFt	150.00
	Add for Finish, Electricity and Erection	Room	9,500.00
.3	X-RAY		
	Deep Therapy		
	Example: Room 10' x 15'	Room	12,000.00
		or SqFt	80.00
	Radiography or Fluoroscopy		
	Example: Room 10' x 15'	Room	5,100.00
		or SqFt	34.00
.4	RADIO FREQUENCY SHIELDING		
	Example: Room 10' x 10'	Room	30,000.00
		or SqFt	3,000.00
	See 1104.10 for X-Ray Equipment		
1312.0	**SAUNAS (M) (Carpenters)**		
	Examples (includes Heaters):		
	Room 5' x 6' Cedar (5 K.W. Unit)	Room	4,500.00
		or SqFt	150.00
	Room 6' x 6' Cedar (6 K.W. Unit)	Room	5,400.00
		or SqFt	150.00
	Room 6' x 7' Cedar (7.5 K.W. Unit)	Room	5,400.00
		or SqFt	130.00
	Add for Electrical Connections and Ventilating		750.00

1313.0 SCALES (NO PORTABLE) (Ironworkers)

.1 FLOOR SCALES (HEAVY DUTY INDUSTRIAL)

Size	Capacity	Labor	Material	Concrete Work
46" x 38"	800 lb.	410.00	4,100.00	-
48" x 48"	1,350 lb.	440.00	4,350.00	-
60" x 48"	1,600 lb.	620.00	4,600.00	-
72" x 48"	2,600 lb.	630.00	5,350.00	-
76" x 54"	6,000 lb.	1,100.00	5,650.00	-
84" x 60"	10,000 lb.	1,850.00	7,500.00	-

.2 TRUCK SCALES - FULL WITH WEIGH BEAM

Truck (4-Section) -

Size	Capacity	Labor	Material	Concrete Work
45" x 10"	50 Ton	2,600.00	20,400.00	20,000.00
50" x 10"	50 Ton	2,900.00	21,000.00	20,500.00
60" x 10"	50 Ton	3,300.00	22,000.00	21,000.00
70" x 10"	50 Ton	3,900.00	23,000.00	22,400.00
Add for 60-Ton Capacity			3,300.00	

Axle Load -

Size	Capacity	Labor	Material	Concrete Work
8" x 10"	20 Ton	1,850.00	9,300.00	6,700.00
8" x 10"	30 Ton	2,350.00	10,200.00	6,800.00
10" x 10"	20 Ton	2,500.00	10,700.00	7,300.00
10" x 10"	30 Ton	2,900.00	11,900.00	7,400.00
Add to Above for Printer and Dial			2,600.00	

.3 CRANE SCALES -

Capacity	Labor	Material	Concrete Work
1 Ton	140.00	2,400.00	-
5 Ton	270.00	2,600.00	-

1314.0 SIGNS (Special Exterior) (L&M) (Sheet Metal Workers)

1315.0 SKYDOMES (L&M) (Carpenters)

EXTRUDED ALUMINUM FRAME, WIRE GLASS
EXTERIOR AND FIBERGLASS SUB CEILING

	UNIT	COST
12' x 18' Ridge Type	Each	7,900.00
	or SqFt	37.00
18' x 18' Pyramid Type	Each	15,200.00
	or SqFt	47.00
36' Diameter Geometric Domes	Each	62,000.00
	or SqFt	60.00
18' Diameter Revolving Domes	Each	16,000.00
	or SqFt	58.00

1316.0 SOUND INSULATED ROOMS (Including Vibration Control) (M) (Carpenters)

	UNIT	COST
.1 ANECHOIC ROOMS (ISOLATORS, PANELS AND WEDGES)	Room	117,000.00
Example: Room 30' x 30'	or SqFt	130.00
.2 AUDIOMETRIC ROOM (PREFABRICATED PANELS)		
Example: Research Room - 5' x 8'	Room	10,400.00
	or SqFt	260.00
Example: Medical Practice Room - 5' x 8'	Room	19,200.00
	or SqFt	480.00

1317.0 SOUND AND VIBRATION CONTROL SYSTEMS

1318.0 SWIMMING POOLS

.1 POOLS

Deep Pools - 3 to 9 feet
(Incl. Filters, Skimmers, Lights, Clng. Equip.)
Rectangular or L Shape:

	UNIT	COST
Residential - 1,000 SqFt	Each	37,000.00
	or SqFt	37.00
Motels and Apartments - 2,000 SqFt	Each	86,000.00
	or SqFt	43.00
Municipal - 5,000 SqFt	Each	250,000.00
	or SqFt	50.00
Add for Diving Stand - Steel - 1 Meter	Each	3,000.00
Steel - 3 Meter	Each	4,400.00
Add for Diving Boards - 12' Fiberglass	Each	820.00
16' Aluminum	Each	1,500.00
Add for Slides - 6' Fiberglass	Each	780.00
12' Fiberglass	Each	1,200.00
Add for Ladders - 10'	Each	300.00
Add for Lifeguard Chairs	Each	1,000.00
Add for Covers	Each	730.00

Shallow Pools - 3 to 5 feet -
Gunite with Plaster Finish

	UNIT	COST
Residential - 1,000 SqFt	Each	28,000.00
	or SqFt	28.00
Motels and Apartments - 2,000 SqFt	Each	72,000.00
	or SqFt	36.00
Add for Water, Drainage & Power to Pools	Each	-
Deduct for Vinyl Lined Pools	SqFt	9.00
Add for Free Form Pools	Each	5%
Add for Heaters	Each	3,200.00
Add for Automatic Cleaners	Each	3,000.00
Wading Pools - 12' x 12' or 9' Round	Each	4,500.00
	or SqFt	32.00

	UNIT	COST
.2 DECKS - CONCRETE - Average 10' around Pool	Each	3,600.00
and Average 1,000 SqFt	or SqFt	3.60
Add for Tile	SqFt	7.25
Add for Brick	SqFt	6.40
.3 FENCES - 4' ALUMINUM OR GALVANIZED	LnFt	1,020.00

1319.0 TANKS

.1 STEEL - GROUND LEVEL (NO FOUNDATION)

	UNIT	COST
150,000 Gals	Each	128,000.00
500,000 Gals	Each	215,000.00
1,000,000 Gals	Each	355,000.00
2,000,000 Gals	Each	560,000.00

.2 STEEL - ELEVATED

	UNIT	COST
100,000 Gals	Each	275,000.00
250,000 Gals	Each	370,000.00
500,000 Gals	Each	600,000.00
1,000,000 Gals	Each	1,300,000.00

1320.0 WOOD DECKS - WITH RAILS

	UNIT	COST
CEDAR	SqFt	13.70
TREATED FIR	SqFt	12.30
Add for Seats	LnFt	3.50
Add for Stairs	Riser	46.00

		PAGE
1401.0	**ELEVATORS (Passenger, Freight & Service) (CSI 14200)**	**14-2**
.1	ELECTRIC (CABLE OR TRACTION)	14-2
.11	Passenger	14-2
.12	Freight	14-2
.2	OIL HYDRAULIC	14-2
.21	Passenger	14-2
.22	Freight	14-2
.3	RESIDENTIAL (WINDING DRUM) (CSI 14235)	14-2
1402.0	**MOVING STAIRS AND WALKS**	**14-3**
.1	MOVING STAIRS (ESCALATORS) (CSI 14200)	14-3
.2	MOVING WALKS (CSI 14320)	14-3
1403.0	**DUMBWAITERS (CSI 14100)**	**14-3**
.1	ELECTRIC	14-3
.2	HAND OPERATED	14-3
1404.0	**LIFTS (CSI 14400)**	**14-3**
.1	WHEELCHAIR LIFTS	14-3
.2	STAIR LIFTS	14-3
.3	VEHICLE - HYDRAULIC LIFTS	14-3
.4	MATERIAL HANDLING LIFTS	14-3
.5	STAGE LIFTS	14-3
1405.0	**PNEUMATIC TUBE SYSTEMS (CSI 14580)**	**14-3**
1406.0	**MATERIAL HANDLING (CSI 14500)**	**14-3**
1407.0	**TURNTABLES (CSI 14700)**	**14-3**
1408.0	**POWERED SCAFFOLDING (CSI 14840)**	**14-3**

1401.0   ELEVATORS (L&M) (Elevator Constructors & Ironworkers)

.1   ELECTRIC (CABLE/ TRACTION)

.11   Passenger - Std Specifications

	Speed to F.P.M.	Capacity and Cost			
		2,000#	3,000#	4,000#	10,000#
Geared A.C. Rheostatic:					
Selective Collective Operation					
3 Openings	125	66,000	-	-	-
4 Openings	125	69,000	-	-	-
5 Openings	125	72,000	-	-	-
Add per Floor	-	5,000	-	-	-
Geared Variable Voltage:					
Selective Collective Operation					
8 Openings	250	121,000	124,000	127,000	-
	350	125,000	129,000	130,000	-
10 Openings	250	124,000	128,000	132,000	-
	350	128,000	126,000	135,000	-
Add per Floor	-	6,500	7,000	7,500	-
Group Supervisory					
8 Openings	250	-	134,000	136,000	-
	350	-	142,000	141,000	-
10 Openings	250	-	134,000	139,000	-
	350	-	141,000	142,000	-
Add per Floor	-	-	7,000	7,500	-
Gearless Variable Voltage:					
Group Supervisory					
8 Openings	500	-	167,000	163,000	-
	800	-	180,000	176,000	-
	1000	-	190,000	190,000	-
10 Openings	500	-	170,000	152,000	-
	800	-	185,000	189,000	-
	1000	-	200,000	190,000	-
Add per Floor	-	-	6,800	7,000	-
Add per Fl. of Express Zone	-	6,300	6,700	-	-

.12   Freight

	Speed to F.P.M.	2,000#	3,000#	4,000#	10,000#
3 Openings	100	-	-	88,000	93,000
4 Openings	100	-	-	93,000	98,000
Add per Floor	-	-	-	19,800	11,600
Add per Power Operated Door	-	-	-	-	7,000

.2   OIL HYDRAULIC

.21   Passenger

	Speed to F.P.M.	2,000#	3,000#	4,000#	10,000#
2 Openings	100	47,000	-	-	-
3 Openings	100	51,000	-	-	-
4 Openings	100	58,000	-	-	-
5 Openings	100	65,000	-	-	-
6 Openings	100	74,000	-	-	-

Deduct for Limited Use Elevator, non-Commercial 8,000# - $5,000

.22   Freight

	Speed to F.P.M.	2,000#	3,000#	4,000#	10,000#
2 Openings	100	-	-	55,000	67,000
3 Openings	100	-	-	64,000	71,000
4 Openings	100	-	-	66,000	75,000

Add per Power Operated Door - $11,000

.3   RESIDENTIAL (WINDING DRUM)

   2 Openings - Speed to 36 F.P.M., 700# Capacity - $19,000
      Speed to 30 F.P.M., 410# Capacity - $14,500
   Add for 3 Openings - $3,300

   Add for ADA Updating -$12,000
   Add for Updating Generators, Controllers & Operating Equipment -$55,000
   Add to 1401.1 and 1401.2 Above for Custom Specifications - 10%
   Add to 1401.2 for Shaft Work, Electric Power & Mechanical - approx. $53,000
   Add for Inspections and Electric Work - $1,400

1402.0 MOVING STAIRS & WALKS (L&M) (Elevator Constructors & Ironworkers)

.1 MOVING STAIRS (Escalators)

	UNIT	CAPACITY	COST
16' Floor Height			
Metal Balustrade 32" Wide			111,000
48"			122,000
Glass Balustrade 32" Wide			121,000
48"			132,000
Add or Deduct per Foot Floor Height			2,700
Add for Over 20 Feet Floor Height			25%

.2 MOVING WALKS - Belt Type 32"

	UNIT	CAPACITY	COST
MOVING WALKS - Belt Type 32"	LnFt		725
48"	LnFt		790
Pallet Type 32"	LnFt		775
48"	LnFt		850

1403.0 DUMBWAITERS (L&M) (Elevator Constructors)

.1 ELECTRIC

	Speed F.P.M.		COST
Drum Type			
2 Openings - 1 Stop	25 to 75	50# to 500#	13,000
3 Openings - 2 Stops	25 to 75	50# to 500#	16,000
Add per Floor			3,500
Traction Type			
3 Openings - 2 Stops	50 to 300	50# to 500#	17,500
4 Openings - 3 Stops	50 to 300	50# to 500#	19,800
5 Openings - 4 Stops	50 to 300	50# to 500#	24,000
6 Openings - 5 Stops	50 to 300	50# to 500#	26,500
Add per Floor			3,500

.2 HAND OPERATED

	COST
2 Openings - 1 Stop	6,000
3 Openings - 2 Stops	6,600
Add per Floor	900
Add for all Stainless Steel	25%

1404.0 LIFTS (L&M) (Ironworkers & Elevator Constructors)

	UNIT	CAPACITY	COST
.1 WHEELCHAIR LIFTS - 42"	Each	1,050#	16,500
96"	Each	1,400#	20,000
.2 STAIR LIFTS (CLIMBERS)	Each		5,400
.3 VEHICLE - SEMI HYDRAULIC LIFTS - 1 Post	Each	8,000#	6,800
2 Post	Each	10,000#	10,800
2 Post	Each	15,000#	16,500
Add for Fully Hydraulic - 2 Post	Each	10,000#	3,800
.4 MATERIAL HANDLING LIFTS	Each		-
.5 STAGE LIFTS - 12' x 30'	Each		82,000
Add for Permits, Inspections & Elec Hookups			1,400

See 1116 for Loading Dock Lift Equipment

1405.0 PNEUMATIC TUBE SYSTEMS (L&M) (Sheet Metal Workers)

	UNIT	COST
3" ROUND - Single Tube	LnFt	27
Twin Tube	LnFt	36
4" ROUND - Single Tube	LnFt	28
Twin Tube	LnFt	37
4" x 7" OVAL - Single Tube	LnFt	35
Twin Tube	LnFt	48

1406.0 MATERIAL HANDLING (L&M) (Ironworkers & Millwrights) -
1407.0 TURNTABLES (L&M) (Ironworkers & Millwrights) -
1408.0 POWERED SCAFFOLDING (L&M) (Ironworkers) -

Wage Rates (Including Fringes) & Location Modifiers
July 2001-2002

	Metropolitan Area	Plumber Wage Rate		Wage Rate Location Modifier
1.	Akron	*	32.99	106
2.	Albany-Schenectady-Troy	*	32.19	100
3.	Atlanta		29.19	88
4.	Austin	**	24.34	74
5.	Baltimore		32.97	101
6.	Birmingham		24.19	70
7.	Boston		42.50	135
8.	Buffalo-Niagara Falls	*	32.17	102
9.	Charlotte	**	20.55	60
10.	Chicago-Gary		41.10	122
11.	Cincinnati		31.17	98
12.	Cleveland		38.45	120
13.	Columbus	*	34.45	95
14.	Dallas-Fort Worth	**	24.15	75
15.	Dayton	*	30.93	97
16.	Denver-Boulder		27.85	90
17.	Detroit		41.88	125
18.	Flint	*	33.08	102
19.	Grand Rapids	**	31.43	96
20.	Greensboro-West Salem	**	20.45	60
21.	Hartford-New Britain	*	36.26	115
22.	Houston	**	26.92	78
23.	Indianapolis	*	34.81	106
24.	Jacksonville	**	26.31	84
25.	Kansas City		36.44	100
26.	Los Angeles-Long Beach	*	38.78	131
27.	Louisville	**	28.22	87
28.	Memphis	**	25.49	91
29.	Miami	**	28.60	87
30.	Milwaukee	*	35.04	110
31.	Minneapolis-St. Paul		36.37	109
32.	Nashville	**	26.11	76
33.	New Orleans	**	23.49	68
34.	New York		59.84	197
35.	Norfolk-Portsmouth	**	25.85	79
36.	Oklahoma City	**	26.80	82
37.	Omaha-Council Bluffs	*	28.04	86
38.	Orlando	**	27.45	85
39.	Philadelphia	*	42.16	139
40.	Phoenix	**	30.33	95
41.	Pittsburgh	*	35.60	111
42.	Portland	*	37.73	106
43.	Providence-Pawtucket	*	33.68	104
44.	Richmond	**	25.74	78
45.	Rochester	**	34.65	108
46.	Sacramento	*	49.56	164
47.	St. Louis		39.39	105
48.	Salt Lake City-Ogden	**	25.75	78
49.	San Antonio	**	27.26	81
50.	San Diego	*	38.68	120
51.	San Francisco-Oakland-San Jose	*	49.56	164
52.	Seattle-Everett		42.18	133
53.	Springfield-Holyoke-Chicopee	*	33.17	103
54.	Syracuse	*	32.87	97
55.	Tampa-St. Petersburg	**	28.69	85
56.	Toledo	*	35.28	110
57.	Tulsa	**	27.62	82
58.	Tucson	**	31.11	96
59.	Washington D.C.	*	33.31	88
60.	Youngstown-Warren	*	31.67	97
	AVERAGE		**32.71**	

Impact Ratio: Labor 35%, Material 65%
*   Contract Not Settled - Wage Interpolated or Unknown
** Non Signatory or Open Shop Rate

		PAGE
1501.0	**BASE MATERIALS AND METHODS (CSI 15050)**	**15-4**
.1	PIPE AND PIPE FITTINGS	15-4
.2	PIPING SPECIALITIES	15-6
.21	Expansion Joints	15-6
.22	Vacuum Breakers	15-6
.23	Strainers	15-6
.3	MECHANICAL - SUPPORTING ANCHORS AND SEALS	15-6
.4	VALVES	15-6
.5	PUMPS	15-7
.6	VIBRATION ISOLATION	15-7
.7	METERS AND GAGES	15-7
.8	TANKS	15-7
1502.0	**MECHANICAL INSULATION**	**15-7**
.1	INSULATION FOR PIPING	15-7
.2	INSULATION FOR DUCT WORK	15-7
.3	INSULATION FOR BOILERS	15-7
1503.0	**FIRE PROTECTION EQUIPMENT (CSI 15300)**	**15-8**
.1	AUTOMATIC SPRINKLER EQUIPMENT	15-8
.2	CARBON DIOXIDE EQUIPMENT	15-8
.3	STANDPIPE AND FIRE HOSE EQUIPMENT	15-8
1504.0	**PLUMBING (CSI 15400)**	**15-8**
.1	EQUIPMENT	15-8
.2	PACKAGE WASTE, VENT, OR WATER PIPING	15-8
.3	DOMESTIC WATER SOFTENER	15-8
.4	PLUMBING FIXTURES	15-8
.5	RESIDENTIAL PLUMBING	15-8
.6	POOL EQUIPMENT	15-8
.7	FOUNTAIN PIPING	15-8
1505.0	**HEAT GENERATION (CSI 15500)**	**15-9**
.1	FUEL HANDLING EQUIPMENT	15-9
.2	ASH REMOVAL SYSTEM	15-9
.3	LINED BREECHINGS	15-9
.4	BOILERS	15-9
1506.0	**REFRIGERATION (CSI 15600)**	**15-9**
.1	REFRIGERANT COMPRESSORS	15-9
.2	CONDENSERS	15-9
.3	CHILLERS	15-9
.4	COOLING TOWERS	15-9
1507.0	**HEAT TRANSFERS (CSI 15750)**	**15-9**
.1	HOT WATER SPECIALITIES	15-9
.2	CONDENSATE PUMPS AND RECEIVER SETS	15-9
.3	HEAT EXCHANGERS	15-9
.4	TERMINAL UNITS	15-9
.5	COILS	15-10
.6	UNIT HEATERS	15-10
.7	PACKAGED HEATING AND COOLING	15-10
.8	STEAM SPECIALTIES	15-10

PAGE

1508.0   AIR HANDLING AND DISTRIBUTION (CSI 15850 AND 15880)   **15-10**

.1   FURNACE   15-10
.2   FANS   15-10
.3   DUCT WORK   15-10
.4   DUCT ACCESSORIES   15-10
.5   RESIDENTIAL HEATING AND AIR CONDITIONING   15-10

1509.0   SEWAGE DRAINAGE (CSI 02700)   **15-10**
.1   SEWAGE EJECTORS   15-10
.2   GREASE INTERCEPTORS   15-10
.3   LIFT STATIONS - STEEL, CONCRETE OR FIBERGLASS   15-10
.4   SEPTIC TANKS AND DRAIN FIELD   15-10
.5   SEWAGE TREATMENT EQUIPMENT - STEEL OR CONCRETE   15-10
.6   AERATION EQUIPMENT - STEEL   15-10
.7   SLUDGE DIGESTION   15-10

1510.0   CONTROLS AND INSTRUMENTATION (CSI 15950)   **15-10**
.1   ELECTRICAL AND INTERLOCKS   15-10
.2   IDENTIFICATION   15-10
.3   CONTROL PIPING, TUBING AND WIRING   15-10
.4   CONTROL AIR COMPRESSOR AND DRYER   15-10
.5   CONTROL PANELS   15-10
.6   INSTRUMENT PANELBOARD   15-10
.7   PRIMARY CONTROL DEVICES   15-10
.8   RECORDING DEVICES   15-10
.9   ALARM DEVICES   15-10

1511.0   TESTING, ADJUSTING AND BALANCING (CSI 15990)   **15-10**

Costs of mechanical work are priced as total contractor's costs with an overhead and fee of 20% included. Not included are scaffold, hoisting, temporary heat, enclosures and bonding. Included are 35% taxes and insurance on labor, equipment, tools and 5% sales tax.

1501.0 BASE MATERIALS AND METHODS

		UNIT	COST
.1	PIPE AND PIPE FITTINGS (Including Fittings)		
	Steel - Black - Threaded 3/4"	LnFt	2.85
	1"	LnFt	3.30
	1-1/2"	LnFt	4.70
	2"	LnFt	9.00
	2-1/2"	LnFt	12.50
	3"	LnFt	17.30
	4"	LnFt	21.70
	5"	LnFt	25.00
	6"	LnFt	32.50
	8"	LnFt	47.50
	Steel - Galvanized Schedule 40 3/4"	LnFt	6.00
	1"	LnFt	7.60
	1-1/4"	LnFt	8.80
	1-1/2"	LnFt	10.40
	2"	LnFt	13.50
	3"	LnFt	21.75
	4"	LnFt	29.20
	5"	LnFt	35.30
	6"	LnFt	48.00
	8"	LnFt	54.00
	Deduct for Black Iron	LnFt	5%
	Deduct for Underground	LnFt	10%
	Cast Iron - Hub and Heavy Duty 2"	LnFt	13.80
	3"	LnFt	14.50
	4"	LnFt	20.00
	5"	LnFt	23.80
	6"	LnFt	32.60
	8"	LnFt	38.50
	Add for Line Pipe	LnFt	4.25
	Deduct for Hubless Type	LnFt	10%
	Deduct for Standard	LnFt	15%
	P.V.C. - Schedule 40 1/2"	LnFt	4.25
	3/4"	LnFt	4.70
	1"	LnFt	5.20
	1-1/4"	LnFt	5.50
	1-1/2"	LnFt	6.20
	2"	LnFt	6.75
	3"	LnFt	7.80
	4"	LnFt	13.30
	6"	LnFt	23.00
	Deduct for Underground	LnFt	10%
	P.V.C. - Type L 3/8"	LnFt	3.15
	1/2"	LnFt	4.20
	3/4"	LnFt	4.80
	1"	LnFt	5.60
	1-1/4"	LnFt	6.20
	1-1/2"	LnFt	7.00
	2"	LnFt	7.80
	3"	LnFt	9.00
	4"	LnFt	13.50
	6"	LnFt	26.00
	Add for Type K (Underground)	LnFt	20%
	Deduct for Type M (Residential)	LnFt	5%

1501.0 BASE MATERIALS AND METHODS (Cont'd...)

.1 PIPE AND PIPE FITTINGS (Cont'd...)

	UNIT	COST
Brass - Threadless 1/2"	LnFt	7.45
3/4"	LnFt	8.20
1"	LnFt	10.30
1-1/4"	LnFt	13.00
1-1/2"	LnFt	15.50
2"	LnFt	18.80
3"	LnFt	26.00
4"	LnFt	34.00
Add for Threaded	LnFt	30%
Stainless Steel Schedule 40 1/4"	LnFt	7.50
1/2"	LnFt	10.70
3/4"	LnFt	14.10
1"	LnFt	16.50
2"	LnFt	22.80
4"	LnFt	56.00
Add for Type 316	LnFt	20%
Fibre Glass - 1"	LnFt	15.00
1-1/2"	LnFt	17.20
2"	LnFt	19.50
3"	LnFt	22.40
4"	LnFt	25.80
5"	LnFt	33.00
6"	LnFt	50.00
8"	LnFt	58.00

Utility Piping -

	UNIT	Reinforced Concrete	Vitrified Clay	Bituminous Coated Corrugated	Smooth Paved Corrugated	Ductile Iron
6"	LnFt	$6.00	$6.30	$8.30	$9.00	$14.30
8"	LnFt	6.40	8.80	8.80	9.50	20.70
1"	LnFt	8.20	11.50	10.30	11.40	23.80
12"	LnFt	10.00	13.70	12.50	14.00	33.20
15"	LnFt	12.00	20.00	15.50	16.60	43.00
18"	LnFt	15.50	29.80	19.30	28.70	56.00
21"	LnFt	19.50	36.30	22.70	25.00	67.00
24"	LnFt	24.50	53.00	25.00	29.60	79.00
27"	LnFt	33.00	74.00	30.30	35.20	-
30"	LnFt	41.40	98.00	32.00	40.50	-
36"	LnFt	54.50	134.00	35.30	49.50	-
42"	LnFt	71.50	-	41.00	57.00	-
48"	LnFt	90.00	-	56.00	64.00	-
54"	LnFt	105.00	-	72.25	85.00	-
60"	LnFt	125.00	-	90.00	108.00	-

Add for Trenching and Backfill - See Division 0202.0
See 0205 for Site Drainage Items

1501.0 BASE MATERIALS AND METHODS, Cont'd...

.2 PIPING SPECIALTIES

	UNIT	COST
.21 Expansion Joints - Copper 1/2"	Each	$52.00
3/4"	Each	78.00
1"	Each	124.00
1-1/2"	Each	260.00
2"	Each	393.00
3"	Each	725.00
4"	Each	790.00
.22 Vacuum Breakers - 1/2"	Each	33.00
3/4"	Each	38.50
1"	Each	52.00
1-1/2"	Each	71.50

	UNIT	IRON BODY	BRONZE BODY
.23 Strainers - Screwed Ends - 1/2"	Each	28.00	36.00
3/4"	Each	32.00	41.50
1"	Each	38.50	56.00
1-1/2"	Each	53.00	89.00
2"	Each	85.00	145.00
3"	Each	145.00	280.00
4"	Each	395.00	10,300.00
Strainers - Flanged Ends - 2"	Each	135.00	375.00
3"	Each	238.00	760.00
4"	Each	385.00	1,330.00
5"	Each	610.00	1,800.00
6"	Each	790.00	2,500.00

.3 MECHANICAL - SUPPORTING ANCHORS & SEALS

.4 VALVES

Bronze	UNIT	150 psi Gate	150 psi Check	150 psi Globe	Relief	300 psi to 75 Reducing
Body - 1/2"	Each	$26.00	$23.00	$30.00	$44.00	$62.00
3/4"	Each	32.00	26.00	40.00	53.00	73.00
1"	Each	38.00	30.00	53.00	78.00	108.00
1-1/2"	Each	60.00	49.00	84.00	250.00	240.00
2"	Each	78.00	74.00	120.00	275.00	360.00
2-1/2"	Each	138.00	420.00	205.00	-	-
3"	Each	190.00	160.00	310.00	-	-

Iron	UNIT	125 psi Gate	125 psi Check	125 psi Globe
Body - 2"	Each	$220.00	$175.00	$315.00
2-1/2"	Each	230.00	178.00	365.00
3"	Each	260.00	182.00	435.00
4"	Each	380.00	330.00	580.00
6"	Each	550.00	350.00	735.00

	UNIT	Gate	Check	Globe
Steel - 2"	Each	$750.00	$725.00	$930.00
2-1/2"	Each	1,160.00	970.00	1,370.00
3"	Each	1,170.00	990.00	1,400.00
4"	Each	1,400.00	1,390.00	2,020.00

	UNIT	Ball	Foot
Plastic - 1/2"	Each	$21.50	$50.50
3/4"	Each	26.00	56.00
1"	Each	30.00	73.00
1-1/2"	Each	51.00	118.00
2"	Each	58.00	140.00

1501.0	BASE MATERIALS AND METHODS, Cont'd...		UNIT	COST
.5	PUMPS			
	Condensate Return - Simplex		Each	$ 1,650.00
	Duplex		Each	4,300.00
	Utility Water Pump -			
	Multi - Two Stage 3" & 4" 75 HP		Each	16,500.00
	Four Stage 3" & 4" 150 HP		Each	30,500.00
	Single - End Suction 1" x 2" 3 HP		Each	4,200.00
	End Suction 2" x 3" 15 HP		Each	5,400.00
	Condenser Water Pump, Pressure - 100 GPM		Each	14,500.00
	300 GPM		Each	19,200.00
	1200 GPM		Each	42,200.00
	2000 GPM		Each	52,000.00
	Submersible Pump 2"		Each	600.00
	Pedestal Pump		Each	185.00
.6	VIBRATION ISOLATION 5" long		Each	310.00
	6" long		Each	590.00
.7	METERS AND GAGES			
	Disc Type - 3/4"		Each	150.00
	1"		Each	185.00
	2"		Each	510.00
	Compound Type - 3"		Each	1,780.00
	4"		Each	2,800.00
.8	TANKS			
	Fuel Storage - Steel - 1,000 Gallons		Each	1,720.00
	2,000		Each	2,400.00
	5,000		Each	5,500.00
	10,000		Each	7,900.00
	Fuel Storage - Fiberglass - 2,000 Gallons		Each	2,900.00
	5,000		Each	5,100.00
	10,000		Each	8,400.00
	Hot Water Storage - 80 Gallons		Each	350.00
	140		Each	530.00
	200		Each	730.00
	400		Each	1,120.00
	600		Each	1,700.00
	Liquid Expansion - 30 Gallons		Each	310.00
	60		Each	500.00
	100		Each	680.00

1502.0 MECHANICAL INSULATION

.1	INSULATION FOR PIPING	Unit	3/8" Foam	1" Wall Fiberglass	2" Wall Fiberglass
	Interior - Pipe 1/2"	LnFt	3.20	3.90	5.50
	3/4"	LnFt	3.30	4.15	5.90
	1"	LnFt	3.60	4.30	6.00
	1-1/4"	LnFt	3.80	4.65	6.25
	1-1/2"	LnFt	4.00	4.90	6.60
	2"	LnFt	4.40	5.25	7.30
	2-1/2"	LnFt	4.75	5.60	8.00
	3"	LnFt	5.10	6.10	9.20
	6"	LnFt	7.25	7.70	12.50
	8"	LnFt	8.90	11.20	14.00

Exterior - w/Aluminum Jacket - Add 2.00/ LnFt

.2	INSULATION FOR DUCT WORK	Unit	1"	1-1/2"	2"
	Rigid w/ Vapor Barrier	SqFt	2.05	2.85	2.30
	Blanket	SqFt	4.30	3.15	2.55
.3	INSULATION FOR BOILERS				
	Fiberglass - 2"	SqFt	-	-	9.90
	Calcium Silicate - 2" with C.F.	SqFt	-	-	15.00

		UNIT	COST
1503.0	**FIRE PROTECTION EQUIPMENT**		
.1	AUTOMATIC SPRINKLER EQUIPMENT - Wet System	SqFt	1.80
		(or) Head	175.00
	Dry System	SqFt	2.10
		(or) Head	188.00
	Add for Pendant Type Sprinklers	SqFt	.50
	Add for Deluge System	SqFt	1.75
.2	CARBON DIOXIDE EQUIPMENT - Cylinder	Each	840.00
	Detector & Nozzle	Each	280.00
.3	STANDPIPE AND FIRE HOSE EQUIPMENT		
	Standpipe - Exposed 6"	Each	525.00
	Concealed 6"	Each	670.00
	Cabinets - 30" x 30" Aluminum	Each	280.00
	Hose and Nozzle - 1 1/2"	LnFt	2.50
	2 1/2"	LnFt	3.00
.4	HALON SYSTEM (COMPUTER ROOM)	Each	3,500.00
1504.0	**PLUMBING**		
.1	EQUIPMENT	-	-
.2	PACKAGE WASTE, VENT OR WATER PIPING	Each	390.00
.3	DOMESTIC WATER SOFTENER	Each	740.00
.4	PLUMBING FIXTURES (Drain, Waste and Vent Not Included - See 1504.5)		
	Porcelain Enameled (unless otherwise noted) **with accessories**		
	Bathtub - 5' - Cast Iron - Enameled	Each	980.00
	Steel - Enameled	Each	495.00
	Fiberglass	Each	515.00
	5' x 5' - Cast Iron - Enameled	Each	1,290.00
	Bathtub/ Whirlpool - Fiberglass 6'	Each	3,550.00
	Bidet	Each	850.00
	Drinking Fountain - Enameled Wall Hung	Each	540.00
	SS Recessed	Each	650.00
	Garbage Disposal - 1/2 Horsepower	Each	150.00
	Lavatories - 19" x 17"- Oval	Each	255.00
	20" x 18" - Rectangular	Each	310.00
	27" x 23" - Oval Pedestal	Each	660.00
	18" - SS Bowl	Each	340.00
	Add for Handicap Lavatory	Each	165.00
	Shower & Tub - Combined - Fiberglass	Each	1,110.00
	Shower/Stall - Fiberglass - 32" x 32"	Each	730.00
	Fiberglass - 36" x 36"	Each	970.00
	Sink, Kitchen - 24" x 21" - Single	Each	475.00
	24" x 21" - SS Single	Each	335.00
	33" x 22" - Double	Each	520.00
	32" x 21" - SS Double	Each	390.00
	Sink, Laundry - Enameled	Each	520.00
	Plastic	Each	335.00
	Sink, Service - 24" x 21"	Each	630.00
	Urinals - Wall Hung	Each	560.00
	Stall or Trough 4'	Each	760.00
	Wash Fountain - Circular - Precast Terrazzo	Each	1,840.00
	Semi-circular - Precast Terrazzo	Each	1,800.00
	Water Closets - Floor Mounted - Tank Type	Each	410.00
	One Piece	Each	560.00
	Wall Mounted	Each	600.00
	Add for Colored Fixtures	Each	25%
	Add for Handicap Water Closet	Each	90.00
	See 1106 for Appliances		
.5	RESIDENTIAL PLUMBING		
	Average Domestic Fixture, Rough In and Trim	Each	455.00
	Average Cost of Drain, Waste, and Vent	Each	800.00
	Floor Drains - 4"	Each	415.00
.6	POOL EQUIPMENT - See 1318.0	-	-
.7	FOUNTAIN PIPING		

			UNIT	COST
1505.0	**HEAT GENERATION**			
.1	FUEL HANDLING EQUIPMENT			-
.2	ASH REMOVAL SYSTEM - Electric		Each	7,800.00
		Hand	Each	4,850.00
.3	LINED BREECHINGS			
.4	BOILERS			-
	Package Type - 1,000 Lbs per Hour		Each	24,000.00
	Steam & Hot Water - 2,000 Lbs per Hour		Each	30,000.00
	5,000 Lbs per Hour		Each	50,000.00
	10,000 Lbs per Hour		Each	55,000.00
	15,000 Lbs per Hour		Each	78,000.00
	20,000 Lbs per Hour		Each	84,000.00
	25,000 Lbs per Hour		Each	113,000.00
	Cast Iron - 100 MBH		Each	2,100.00
	Hot Water or Steam - 200 MBH		Each	3,700.00
	300 MBH		Each	4,600.00
	600 MBH		Each	7,800.00
	1,000 MBH		Each	13,000.00
	Hot Water Heaters - 20 GPM		Each	540.00
	30 GPM		Each	725.00
	40 GPM		Each	780.00
	50 GPM		Each	1,020.00
1506.0	**REFRIGERATION**			
.1	REFRIGERANT COMPRESSORS			-
.2	CONDENSERS - Air Cooled - 10 Ton		Each	12,400.00
	20 Ton		Each	15,000.00
	30 Ton		Each	21,600.00
	40 Ton		Each	26,000.00
	60 Ton		Each	37,000.00
	80 Ton		Each	44,300.00
	Add for Water Cooled		Each	10%
.3	CHILLERS - Hermetic Centrifugal - 100 Ton		Each	48,500.00
	150 Ton		Each	51,000.00
	300 Ton		Each	71,000.00
	Add for Absorption Type		Each	15%
.4	COOLING TOWERS - 100 Ton		Each	6,700.00
	150 Ton		Each	12,400.00
	300 Ton		Each	33,500.00
1507.0	**HEAT TRANSFERS**			
.1	HOT WATER SPECIALTIES			-
.2	CONDENSATE PUMPS AND RECEIVER SETS			
	Outdoor - Air Cooled - 30,000 BTU		Each	1,680.00
	50,000 BTU		Each	2,180.00
	Indoor - 30,000 BTU		Each	860.00
	50,000 BTU		Each	1,190.00
.3	HEAT EXCHANGERS - 100 GPM		Each	970.00
	200 GPM		Each	1,020.00
	300 GPM		Each	2,400.00
	600 GPM		Each	4,750.00
	1,000 GPM		Each	6,400.00
.4	TERMINAL UNITS			
	Base Boards - Cast Iron Radiant - 7 1/4"		LnFt	20.00
	9 3/4"		LnFt	21.40
	Tube with Cover - 7 1/4" - 3/4"		LnFt	11.00
	7 1/4" - 1"		LnFt	12.00
	Tube with Cover - 9 3/4" - 3/4"		LnFt	13.60
	9 3/4" - 1"		LnFt	16.70
	14" Fin Tube - Steel		LnFt	20.50
	Copper		LnFt	25.00
	Convectors		Each	240.00
	Radiators		SqFt	7.25

		UNIT	COST
1507.0	**HEAT TRANSFERS, Cont'd...**	-	-
.5	COILS	-	-
.6	UNIT HEATERS - 100,000 BTU	Each	620.00
	150,000 BTU	Each	840.00
	200,000 BTU	Each	1,000.00
.7	PACKAGED HEATING AND COOLING	-	1,900.00
.8	STEAM SPECIALTIES	-	430.00
1508.0	**AIR HANDLING AND DISTRIBUTION**		
.1	FURNACE - Duct - 100 MBH	Each	1,110.00
	200 MBH	Each	1,580.00
	300 MBH	Each	2,800.00
.2	FANS - 1,000 CFM	Each	115.00
	5,000 CFM	Each	500.00
	10,000 CFM	Each	825.00
.3	DUCT WORK - Galvanized - 26 Ga	Lb	2.70
	24 Ga	Lb	2.90
	22 Ga	Lb	3.30
	20 Ga	Lb	3.60
.4	DUCT ACCESSORIES		
	Constant Volume Duct	Each	490.00
	Reheat Constant Volume Duct	Each	600.00
	Diffusers & Registers - Sidewalk	SqFt	25.00
	Ceiling	SqFt	43.00
.5	RESIDENTIAL HEATING AND AIR CONDITIONING		
	Furnace	Each	1,250.00
	Central Air Conditioning	Each	600.00
	Humidifier	Each	290.00
	Air Cleaner	Each	475.00
	Thermostat	Each	120.00
	Attic Fan	Each	280.00
	Ventilating Damper	Each	105.00
	6" Flue	Each	320.00
.6	COMMERCIAL AIR CONDITIONING		
	Average	Ton	2,050.00
	High	Ton	3,100.00
	Low	Ton	1,030.00
1509.0	**SEWAGE DRAINAGE**		
.1	SEWAGE EJECTORS 100 GPM	Each	6,600.00
.2	GREASE INTERCEPTORS - 200 GPM	Each	7,800.00
	300 GPM	Each	14,200.00
.3	LIFT STATIONS - STEEL, CONCRETE OR FIBERGLASS		
	WITH GENERATOR - 100 GPM	Each	67,000.00
	200 GPM	Each	113,000.00
	500 GPM	Each	154,000.00
.4	SEPTIC TANKS AND DRAIN FIELD		
	Concrete Tanks with Excav. - 1,000 Gallons	Each	1,350.00
	2,000	Each	2,800.00
	5,000	Each	6,200.00
	10,000	Each	11,500.00
	Drain Fields	LnFt	4.10
.5	SEWAGE TREATMENT EQUIPMENT - STEEL OR CONCRETE		
	5,000 Gallons	GPD	8.50
	50,000	GPD	3.10
	100,000	GPD	2.90
.6	AERATION EQUIPMENT - STEEL - 1,000 Gallons	Each	15,500.00
	CAPACITY PER DAY 2,000	Each	22,600.00
	5,000	Each	31,000.00
	10,000	Each	37,500.00
.7	SLUDGE DIGESTION		-
1510.0	**CONTROLS AND INSTRUMENTATION** - No cost for items listed on Index		-
1511.0	**TESTING, ADJUSTING AND BALANCING**		

Wage Rates (Including Fringes) & Location Modifiers
July 2001-2002

	Metropolitan Area		Electrician Wage Rate	Wage Rate Location Modifier
1.	Akron	*	34.92	104
2.	Albany-Schenectady-Troy	*	33.00	104
3.	Atlanta		28.75	87
4.	Austin	**	27.68	83
5.	Baltimore	*	33.49	99
6.	Birmingham		23.37	72
7.	Boston		45.41	140
8.	Buffalo-Niagara Falls		37.64	111
9.	Charlotte	**	21.95	63
10.	Chicago-Gary	*	44.79	130
11.	Cincinnati	*	27.84	85
12.	Cleveland		47.18	112
13.	Columbus	**	34.21	98
14.	Dallas-Fort Worth	*	28.51	84
15.	Dayton	*	30.61	97
16.	Denver-Boulder	*	29.45	89
17.	Detroit		41.49	120
18.	Flint	*	33.44	100
19.	Grand Rapids	*	32.97	99
20.	Greensboro-West Salem	**	21.54	62
21.	Hartford-New Britain	*	40.55	118
22.	Houston	**	31.60	92
23.	Indianapolis		35.00	106
24.	Jacksonville	**	27.94	83
25.	Kansas City	*	34.87	103
26.	Los Angeles-Long Beach		40.18	128
27.	Louisville	**	28.72	87
28.	Memphis	**	29.45	84
29.	Miami	**	30.32	92
30.	Milwaukee	*	37.48	109
31.	Minneapolis-St. Paul		38.01	115
32.	Nashville	**	28.19	68
33.	New Orleans		24.11	74
34.	New York		59.03	167
35.	Norfolk-Portsmouth	**	25.26	76
36.	Oklahoma City	**	25.05	80
37.	Omaha-Council Bluffs	*	29.96	91
38.	Orlando	**	28.17	88
39.	Philadelphia		45.91	135
40.	Phoenix	*	29.65	95
41.	Pittsburgh		36.44	106
42.	Portland	*	40.45	110
43.	Providence-Pawtucket	*	36.05	109
44.	Richmond	**	25.72	76
45.	Rochester	*	33.83	105
46.	Sacramento	*	44.78	140
47.	St. Louis		41.37	124
48.	Salt Lake City-Ogden	**	29.80	89
49.	San Antonio	**	29.91	82
50.	San Diego	*	42.66	125
51.	San Francisco-Oakland-San Jose	*	49.19	140
52.	Seattle-Everett		37.60	113
53.	Springfield-Holyoke-Chicopee	*	34.19	103
54.	Syracuse	*	36.36	99
55.	Tampa-St. Petersburg	**	29.30	88
56.	Toledo	*	35.64	108
57.	Tulsa	**	26.91	80
58.	Tucson	**	31.20	94
59.	Washington D.C.	*	34.12	88
60.	Youngstown-Warren	*	32.02	97
	AVERAGE		**33.92**	

Impact Ratio: Labor 35%, Material 65%

*   Contract Not Settled - Wage Interpolated or Unknown
**   Non Signatory or Open Shop Rate

Costs of Electrical Work are priced as total contractors' costs with an overhead and fee of 20% included.

PAGE

1601.0 BASIC MATERIALS & METHODS (CSI 16050) — **16-3**
.1 CONDUITS — 16-3
.2 BOXES, FITTINGS AND SUPPORTS — 16-3
.3 WIRE AND CABLE — 16-3
.4 SWITCHGEAR — 16-3
.5 SWITCHES AND RECEPTACLES — 16-4
.6 TRENCH DUCT AND UNDERFLOOR DUCT — 16-4
.7 MOTOR WIRING — 16-4
.8 MOTOR STARTERS — 16-4
.9 CABLE TRAY — 16-5

1602.0 POWER GENERATION (CSI 16200) — **16-5**

1603.0 HIGH VOLTAGE DISTRIBUTION (CSI 16300) — **16-5**

1604.0 SERVICE AND DISTRIBUTION (CSI 16400) — **16-5**
.1 METERING — 16-5
.2 GROUNDING — 16-5
.3 SERVICE DISCONNECTS — 16-5
.4 DISTRIBUTION SWITCHBOARDS — 16-5
.5 LIGHT AND POWER PANELS — 16-5
.6 TRANSFORMERS — 16-5

1605.0 LIGHTING (CSI 16500) — **16-6**
.1 LIGHT FIXTURES — 16-6
.2 LAMPS — 16-6

1606.0 SPECIAL SYSTEMS (CSI 16600) — **16-7**

1607.0 COMMUNICATION SYSTEMS (CSI 16700) — **16-7**
.1 TELEPHONE EQUIPMENT — 16-7
.2 FIRE ALARM EQUIPMENT — 16-7
.3 SECURITY SYSTEMS — 16-7
.4 CLOCK PROGRAM EQUIPMENT — 16-7
.5 INTER-COMMUNICATION EQUIPMENT — 16-7
.6 SOUND EQUIPMENT — 16-7
.61 Public Address System (Control and Speakers) — 16-7
.62 Nurses and School Call Systems (Panel and Stations) — 16-7
.7 TELEVISION ANTENNA SYSTEMS — 16-8
.8 CLOSED CIRCUIT TELEVISION STATIONS — 16-8
.9 DOCTOR'S REGISTER — 16-8

1608.0 HEATING AND COOLING (CSI 16850) — **16-8**
.1 SNOW MELTING EQUIPMENT — 16-8
.2 ELECTRICAL HEATING EQUIPMENT — 16-8

1609.0 CONTROLS AND INSTRUMENTATION (CSI 16900) — **16-8**
.1 LIGHTING CONTROL EQUIPMENT — 16-8
.2 DIMMING EQUIPMENT — 16-8
.3 TEMPERATURE CONTROL EQUIPMENT — 16-8
1610.0 TESTING — **16-8**
1611.0 RESIDENTIAL WORK — **16-8**

1601.0 BASIC MATERIALS AND METHODS

.1 CONDUITS (Incl Fittings/Supports)

.11 Rigid

	UNIT	1/2"	3/4"	1-1/2"	2"	3"	4"	5"	6"
Aluminum	LnFt	4.00	4.30	5.25	6.90	8.50	14.00	19.30	46.00
Galvanized	LnFt	4.20	4.50	5.65	8.10	9.40	19.00	25.50	52.00
EMT (Thinwall)	LnFt	2.75	3.20	3.50	4.70	5.45	10.00	-	-
PVC	LnFt	2.70	2.80	2.95	3.70	4.70	7.60	-	-
Add for Running Exposed									

.12 Flexible

	UNIT	1/2"	3/4"	1-1/2"	2"	3"	4"	5"	6"
Greenfield	LnFt	2.50	2.70	3.30	4.80	7.00	-	-	-
Sealtite	LnFt	3.20	3.35	4.80	6.70	9.00	21.00	-	-

.13 Ducts

	UNIT	1/2"	3/4"	1-1/2"	2"	3"	4"	5"	6"
Fibre	LnFt	-	-	-	-	2.90	2.95	3.70	4.30
Transite	LnFt	-	-	-	-	2.90	3.40	4.10	4.70

Cost for Conduits includes Labor, Material, Equipment and Fees and Installed in Wood or Steel Construction.

.2 BOXES, FITTINGS AND SUPPORTS (In Above)

.3 WIRE AND CABLE

	UNIT	COST Copper T.H.W.	COST Aluminum Stranded
Wire - 3 - #14 - 600 Volt	LnFt	3.20	3.00
3 - #12 - 600 Volt	LnFt	3.55	3.30
3 - #10 - 600 Volt	LnFt	4.30	3.90
3 - # 8 - 600 Volt	LnFt	5.50	4.60
3 - # 4 - 600 Volt	LnFt	8.70	6.40
3 - # 2 - 600 Volt	LnFt	14.50	7.75
3 - 1/0	LnFt	19.50	11.00
3 - 2/0	LnFt	22.20	13.00
3 - 3/0	LnFt	31.00	14.50
3 - 4/0	LnFt	36.50	16.00
3 - 250 - MCM	LnFt	42.00	19.00
3 - 300 - MCM	LnFt	48.00	19.50
3 - 350 - MCM	LnFt	53.00	23.50
3 - 400 - MCM	LnFt	60.00	25.50
3 - 500 - MCM	LnFt	65.00	28.50

	UNIT	ARMORED COPPER 2 Wire	ARMORED COPPER 3 Wire	NON-METALLIC 2 Wire	NON-METALLIC 3 Wire
Cable - #14 bx - 600 Volt	LnFt	1.70	1.90	1.40	1.75
#12 bx - 600 Volt	LnFt	1.80	2.10	1.55	2.20
#10 bx - 600 Volt	LnFt	2.10	2.50	2.00	2.55
# 8 bx - 600 Volt	LnFt	-	3.25	-	2.90
# 6 bx - 600 Volt	LnFt	-	3.60	-	3.85
# 4 bx - 600 Volt	LnFt	-	4.00	-	4.90
Add for Work over 10'					10%
Add for Work in Masonry Construction					10%
Add for Work in Concrete Construction					20%

.4 SWITCH GEAR

Metal Clad 5 KV	22,000.00
15 KV	26,000.00
Add for Outdoor Type	4,200.00

1601.0   BASIC MATERIALS AND METHODS, Cont'd...

		UNIT	COST
.5	**SWITCHES AND RECEPTACLES**		
	Switches		
	Standard Toggle - 5 P	Each	50.00
	4 W	Each	56.00
	4 WL	Each	70.00
	Mercury - 5 P	Each	82.00
	4 W	Each	150.00
	Safety - 30 Amp	Each	160.00
	60 Amp	Each	205.00
	100 Amp	Each	350.00
	200 Amp	Each	460.00
	Three-way -	Each	82.00
	Add for Dimmer - 600 Watt	Each	77.00
	Add for Dimmer - 1500 Watt	Each	165.00
	Receptacles		
	Single	Each	48.00
	Duplex - GFI	Each	66.00
	Add for Dedicated - Separate Circuit	Each	125.00
	Add for Dedicated - Ground	Each	17.00
	Add for Dedicated - Fourplex	Each	67.00
	Clock, Lamps, etc.	Each	68.00
	Add for Weatherproof	Each	17.00
	Plates - Stainless Steel	Each	8.50
	Brass	Each	13.60
.6	**TRENCH DUCT AND UNDERFLOOR DUCT (Including Fittings)**		
	Underfloor Duct - Average		
	Single Cell - 1 1/4" x 3 1/8"	LnFt	17.00
	1 1/2" x 6 1/2"	LnFt	21.50
	Double Cell - 1 1/4" x 3 1/8"	LnFt	25.20
	1 1/2" x 6 1/2"	LnFt	42.00
	Wireway - Cover Type 2 1/2" x 2 1/2" x 1"	Each	18.10
	2 1/2" x 2 1/2" x 2"	Each	24.00
	2 1/2" x 2 1/2" x 3"	Each	30.00
	4" x 4" x 1"	Each	22.00
	4" x 4" x 2"	Each	28.00
	4" x 4" x 3"	Each	36.00
.7	**MOTOR WIRING (Including Controls)**		
	10 H.P.	Each	3,600.00
	20 H.P.	Each	3,700.00
	30 H.P.	Each	4,200.00
	40 H.P.	Each	4,400.00
	50 H.P.	Each	5,300.00
	100 H.P.	Each	7,000.00
	200 H.P.	Each	10,700.00
.8	**MOTOR STARTERS (600 Volt)**		
	Magnetic - 10 H.P.	Each	380.00
	Non-Reversing - 20 H.P.	Each	540.00
	30 H.P.	Each	790.00
	40 H.P.	Each	910.00
	50 H.P.	Each	1,050.00
	100 H.P.	Each	2,100.00
	200 H.P.	Each	4,100.00
	Add for Reversing Type	Each	80%
	Add for Explosion Proof	Each	40%
	Add for Motor Circuit Protectors	Each	25%
	Add for Fused Switches	Each	10%
	Manual - 1 H.P.	Each	97.00
	3 H.P.	Each	160.00
	7 1/2 H.P.	Each	190.00

1601.0 BASIC MATERIALS AND METHODS, Cont'd...

	UNIT	COST
.9 CABLE TRAY (Aluminum or Galvanized)		
Ladder or Punched - 6" wide	LnFt	15.00
12" wide	LnFt	16.00
18" wide	LnFt	18.00
24" wide	LnFt	20.00

1602.0 POWER GENERATION

Emergency Generators (3-Phase, 4-Wire) (120/208 Volt)

	UNIT	COST
15 k.w.	Each	10,500.00
30 k.w.	Each	16,000.00
40 k.w.	Each	20,000.00
50 k.w.	Each	26,500.00
75 k.w.	Each	31,00.00
100 k.w.	Each	37,000.00
300 k.w.	Each	70,000.00

1603.0 HIGH VOLTAGE DISTRIBUTION - Power Companies

1604.0 SERVICE AND DISTRIBUTION

.1 METERING

.2 GROUNDING (Including Rods and Accessories)

	UNIT	COST
1/2" x 10'	Each	60.00
5/8" x 10'	Each	72.00
3/4" x 10'	Each	85.00
.3 SERVICE DISCONNECTS - 100 amp	Each	300.00
200 amp	Each	450.00
400 amp	Each	880.00
600 amp	Each	1,730.00
.4 DISTRIBUTION SWITCHBOARDS (240 Volt) - 400 amp	Each	3,100.00
800 amp	Each	3,500.00
1600 amp	Each	5,100.00

.5 LIGHT AND POWER PANELS (3-Phase, 4-Wire) (120/208 Volt)

	UNIT	COST
12 circuit	Each	720.00
16 circuit	Each	850.00
20 circuit	Each	1,000.00
24 circuit	Each	1,200.00
30 circuit	Each	1,400.00
36 circuit	Each	1,520.00
Add for Circuit Breakers	Each	225.00
.6 TRANSFORMERS		
Current - 400 amp	Each	3,000.00
600 amp	Each	3,350.00
800 amp	Each	3,800.00
1000 amp	Each	4,200.00

Light and Power	UNIT	SINGLE-PHASE 120/ 240 Volt	3-PHASE 120/ 208 Volt
Dry Type - 3 kva	Each	350.00	510.00
5 kva	Each	520.00	750.00
10 kva	Each	720.00	980.00
15 kva	Each	1,020.00	1,330.00
30 kva	Each	1,600.00	1,800.00
45 kva	Each	2,300.00	2,350.00
75 kva	Each	3,200.00	3,100.00
Oil Filled - 150 kva	Each	-	5,400.00
300 kva	Each	-	10,200.00
500 kva	Each	-	13,300.00

1605.0 LIGHTING

		UNIT	COST
.1	**LIGHT FIXTURES**		
	Fluorescent		
	Surface		
	Industrial - 48" - 1 lamp R.S.	Each	83.00
	H.O.	Each	120.00
	2 lamp R.S.	Each	123.00
	H.O.	Each	153.00
	4 lamp R.S.	Each	135.00
	H.O.	Each	230.00
	96" - 1 lamp H.O.	Each	130.00
	R.S.	Each	230.00
	2 lamp H.O.	Each	115.00
	H.O.	Each	235.00
	Commercial - 12" x 48" - 2 lamp R.S.	Each	130.00
	24" x 24" - 4 lamp R.S.	Each	180.00
	24" x 48" - 4 lamp R.S.	Each	245.00
	48" x 48" - 4 lamp R.S.	Each	360.00
	12" x 48" - 2 lamp R.S.	Each	167.00
	Recessed - 24" x 24" - 4 lamp R.S.	Each	205.00
	24" x 48" - 4 lamp R.S.	Each	230.00
	48" x 48" - 4 lamp R.S.	Each	400.00
	Add for Electronic Ballasts	Each	22.00
	Incandescent - 100 watt	Each	90.00
	200 watt	Each	105.00
	300 Watt	Each	120.00
	Mercury Vapor - 250 watt Single	Each	430.00
	400 watt Single	Each	470.00
	Twin	Each	600.00
	Quartz - 500 watt	Each	160.00
	Sodium - Low Pressure - 35 watt	Each	290.00
	55 watt	Each	425.00
	Explosion Proof		
	Fluorescent- 48" - 2 lamp R.S.	Each	1,025.00
	Incandescent - 48" - 200 watt	Each	260.00
	Mercury Vapor - 48" - 200 watt	Each	640.00
	Emergency - Nickel Cadmium Battery	Each	685.00
	Lead Battery	Each	350.00
	Exit - Standard	Each	105.00
	Exit - Low Voltage - Battery	Each	180.00
	Porcelain	Each	60.00
.2	**LAMPS**		
	Fluorescent - 48" - 40 watt	Each	5.80
	48" - 60 watt	Each	7.50
	96" - 75 watt	Each	10.50
	Incandescent - 25 to 100 watt	Each	2.10
	400 watt	Each	3.25
	Mercury Vapor - 100 watt	Each	42.00
	250 watt	Each	53.00
	400 watt	Each	72.00
	1000 watt	Each	92.00
	Metal Halide - 175 watt	Each	63.00
	400 watt	Each	68.00
	1000 watt	Each	140.00
	Sodium - 300 watt	Each	85.00
	400 watt	Each	92.00
	1000 watt	Each	210.00
	Add for Disposal	Each	1.00

1606.0	SPECIAL SYSTEMS	UNIT	COST
	LIGHTNING PROTECTION		
	Ground Rod and Copper Point - 3/4" to 10"	Each	82.00
	Ground Clamps - 2"	Each	100.00
	Wire - #2	LnFt	1.10
1607.0	**COMMUNICATION SYSTEMS**		
.1	TELEPHONE EQUIPMENT		
	Cabinets 24" x 36"	Each	300.00
	36" x 48"	Each	460.00
	Lines	Each	270.00
	Phones	Each	155.00
	Station	Each	575.00
.2	FIRE ALARM EQUIPMENT		
	Control Panel 10 zones	Each	1,420.00
	20 zones	Each	2,400.00
	Annunciator Panel 10 zones	Each	480.00
	Station	Each	155.00
	Horn	Each	100.00
	Heat Detector	Each	70.00
	Smoke Detector	Each	145.00
.3	SECURITY SYSTEMS		
	Standard Contact	Each	200.00
	Audible Alarm	Each	270.00
	Photo Electric Sensor - 500' Range	Each	430.00
	Monitor Panel - Standard	Each	480.00
	High Security	Each	660.00
	Vibration Sensor	Each	180.00
	Audio Sensor	Each	130.00
	Motion Sensor	Each	310.00
.4	CLOCK PROGRAM EQUIPMENT (Control Panel and Clock with Buzzer)		
	Master Time Clock 10 rm - average	Each	330.00
	20 rm - average	Each	290.00
	30 rm - average	Each	255.00
	Time Clock and Cards	Each	1,080.00
	Add for Electronic and Bells	Each	20%
.5	INTER-COMMUNICATION EQUIPMENT		
	Individual Stations	Each	130.00
	Board - Master	Each	500.00
.6	SOUND EQUIPMENT		
.61	Public Address System (Control and Speakers)		
	3 to 6 Speakers - average each speaker	Each	720.00
	7 to 9 Speakers - average each speaker	Each	660.00
	10 to 20 Speakers - average each speaker	Each	610.00
	Add for Tape / CD	Each	475.00
.62	Nurses & School Call systems (Panel & Stations)		
	10 Stations or Rooms - average each room	Each	660.00
	20 Stations or Rooms - average each room	Each	610.00
	30 Stations or Rooms - average each room	Each	525.00
	Add per Light	Each	65.00

		UNIT	COST
1607.0	**COMMUNICATIONS SYSTEMS, Cont'd...**		
.7	TELEVISION ANTENNA SYSTEMS		
	School Systems - 10 outlets	Each	315.00
	20 outlets	Each	305.00
	30 outlets	Each	280.00
	Apartment Houses - 20 outlets	Each	135.00
	30 outlets	Each	125.00
	50 outlets	Each	120.00
.8	CLOSED CIRCUIT T.V. STATIONS		
	Recorder	Each	3,550.00
	Camera	Each	1,400.00
	Surveillance	Each	1,830.00
.9	DOCTOR'S REGISTER	Each	6,500.00
	Add for Recording Register	Each	5,600.00
1608.0	**HEATING AND COOLING**		
.1	SNOW MELTING EQUIPMENT - 480 volt	SqFt	9.20
.2	ELECTRIC HEATING EQUIPMENT		
	Baseboard - 500 watt	LnFt	34.00
	750 watt	LnFt	33.00
	1000 watt	LnFt	32.00
	Cable Heating	SqFt	15.00
	Unit Heaters - 1500 watt	Each	200.00
	2500 watt	Each	300.00
1609.0	**CONTROLS AND INSTRUMENTATION**		
.1	LIGHTING CONTROL EQUIPMENT		-
.2	DIMMING EQUIPMENT		-
.3	TEMPERATURE CONTROL EQUIPMENT		-
1610.0	**TESTING**		-
1611.0	**RESIDENTIAL WORK**		
	100-Amp Service	Each	770.00
	200-Amp Service - Single Phase	Each	1,200.00
	Three Phase	Each	2,200.00
	Air Conditioner	Each	250.00
	Dimmer - 600 Watt	Each	50.00
	Dishwasher	Each	70.00
	Doorbell	Each	50.00
	Dryer - Electric	Each	100.00
	Dryer or Washer - Gas	Each	70.00
	Duplex Receptacle	Each	25.00
	Dedicated	Each	58.00
	Exhaust Fan - Bath	Each	48.00
	Kitchen	Each	57.00
	Fixture, Average	Each	85.00
	Porcelain	Each	80.00
	Furnace	Each	85.00
	Garbage Disposal	Each	70.00
	Paddle Fan	Each	150.00
	Phone - Data Box	Each	40.00
	Range Circuit	Each	120.00
	Smoke Detectors	Each	57.00
	Switches - Single	Each	25.00
	3-Way	Each	58.00
	Thermostat	Each	42.00
	Water Heater - 20 Amp 120 Volt	Each	110.00
	Whirlpool	Each	105.00

Average Square Foot Costs
2002

Housing (with Wood Frame)

Square Foot Costs do not include Garages, Land, Furnishings, Equipment, Landscaping, Financing, or Architect's Fee. Total Costs, except General Contractor's Fee, are included in each division. See Page 3 for Metropolitan Cost Variation Modifier. See Division 13 for Metal Buildings, Greenhouses, and Air-Supported Structures.

	Division	Houses - 3 Bedrooms with Basement			Houses - 3 Bedrooms without Basement		
		Low Rent	Project	Custom	Low Rent	Project	Custom
1.	General Conditions	$3.60	$3.65	$3.95	$4.70	$4.75	$5.10
2.	Site Work	2.35	2.60	2.90	4.00	4.15	4.20
3.	Concrete	3.10	3.20	3.40	5.60	5.80	6.30
4.	Masonry	4.05	4.30	5.70	5.60	6.00	7.85
5.	Steel	1.00	1.00	1.05	.50	.50	.50
6.	Carpentry	12.70	13.30	14.10	18.00	18.20	19.25
7.	Moisture & Therm. Prot.	3.10	3.10	3.30	4.35	4.50	4.85
8.	Doors, Windows & Glass	6.20	6.45	6.75	8.25	8.45	8.85
9.	Finishes	5.80	6.30	7.65	8.50	8.65	12.20
10.	Specialties	.60	.60	.65	.65	.65	.90
11.	Equipment (Cabinets)	2.25	2.35	3.40	4.00	5.00	6.00
12.	Furnishings	.45	.45	.50	.75	.75	.75
13.	Special Construction	-	-	-	-	-	-
14.	Conveying	-	-	-	-	-	-
15.	Mechanical: Plumbing	4.40	4.45	5.05	6.50	6.50	8.00
	Heating	3.00	3.00	3.20	4.20	4.50	6.35
	Air Cond.	-	-	-	-	-	-
16.	Electrical	2.55	2.60	3.10	3.80	3.80	4.55
	SUB TOTAL	$55.15	$57.35	$64.70	$79.40	$82.20	$95.65
	CONTRACTOR'S SERVICES	5.55	5.75	6.40	7.90	8.20	8.60
	TOTAL AVERAGE Sq.Ft. Cost	$60.70	$65.10	$71.10	$88.30	$90.40	$104.25

Unfinished Basement, Storage and Utility Areas included in Sq.Ft. Costs:				No Basement Utilities In Finished Area		
Average Cost/ Unit	$121,400	$126,200	$176,500	$105,900	$108,120	$141,400
Average Sq.Ft.Size/ Unit	2,000	2,000	2,500	1,200	1,200	1,500
Add Architect's Fee/ Sq.Ft.	$ 4.75	$ 5.20	$ 9.00	$ 4.75	$ 5.20	$ 9.50

	COST	
	Sq. Ft.	Ea. Unit
Add for Finishing Basement Space For Bedrooms or Amusement Rooms, 12' x 20'	$23.00	$5,520
Add for Single Garage (no Interior Finish), 12' x 20'	$22.00	$5,280
Add for Double Garage (no Interior Finish), 20 x 20'	$23.00	$9,200
Add (or Deduct) for Bedrooms, 12' x 12'	$34.00	$4,900
Add per Bathroom with 2 Fixtures, 6' x 8'	$105.00	$5,040
Add per Bathroom with 3 Fixtures, 8' x 8'	$105.00	$6,720
Add for Fireplace	-	$5,500
Add for Central Air Conditioning	-	$4,300
Add for Drain Tiling	-	$1,900
Add for Well	-	$4,400
Add for Sewage System	-	$4,700

Average Square Foot Costs
2002

Housing

Square Foot Costs do not include Land, Furnishings, Equipment, Landscaping and Financing.

	Division	Apartments 2 & 3 Story (1)a	2 & 3 Story (1)b	High Rise (2)	Motels & Hotels High Rise (3)	1 & 2 Story (1)b	High Rise (2)
1.	General Conditions	$ 3.40	$ 3.65	$ 5.10	$ 5.50	$ 3.75	$ 4.75
2.	Site Work	2.10	2.05	3.00	3.00	4.30	3.00
3.	Concrete	1.70	1.70	16.60	12.00	2.05	16.75
4.	Masonry	1.35	8.65	2.25	2.15	8.60	11.10
5.	Steel	1.00	1.00	1.60	1.70	1.25	1.55
6.	Carpentry and Millwork	12.50	11.00	4.15	3.80	10.75	3.30
7.	Moisture Protection	2.85	3.05	1.55	1.55	2.80	1.60
8.	Doors, Windows and Glass	4.75	5.20	5.70	6.50	5.95	7.05
9.	Finishes	8.75	8.80	8.90	8.45	8.70	12.10
10.	Specialties	.65	.65	.70	.70	.85	.85
11.	Equipment	1.45	1.90	1.45	1.45	2.35	2.00
12.	Furnishings	.70	.75	.80	.80	1.75	1.30
13.	Special Construction	-	-	-	-	-	-
14.	Conveying	-	2.00	3.00	3.20	-	3.60
15.	Mechanical:						
	Plumbing	5.30	5.30	5.20	5.10	6.30	8.10
	Heating and Ventilation	4.65	5.00	5.55	5.25	5.00	6.20
	Air Conditioning	1.65	1.70	1.75	1.75	2.15	4.10
	Sprinklers	-	-	1.75	1.60	1.70	1.70
16.	Electrical	5.50	5.50	6.35	6.65	7.75	7.85
	SUB TOTAL	$58.30	$67.90	$75.40	$71.15	$76.00	$96.90
	CONTRACTOR'S SERVICES	5.80	6.80	7.40	7.15	7.50	9.70
	Construction Sq.Ft. Cost	$64.10	$74.70	$82.80	$78.30	$83.50	$106.60
	Add for Architect's Fee	4.60	5.90	6.40	6.70	6.60	7.60
	TOTAL Sq.Ft. Cost	$68.70	$80.60	$89.20	$85.00	$90.10	$114.20
	Average Cost per Unit without Architect's Fee	$60,700	$70,900	$72,700	$62,600	$53,800	$69,300
	Average Cost per Unit with Architect's Fee	$65,100	$76,500	$78,500	$68,000	$58,200	$74,200
	Average Sq.Ft. Size per Unit	950	950	900	800	650	650
	Average Sq.Ft. Living Space	750	750	750	600	450	450

Add for Porches - Wood - $9.00 SqFt
Add for Interior Garages - $8.00 SqFt

(1)a Wall Bearing Wood Frame & Wood or Aluminum Facade
(1)b Wall Bearing Masonry & Wood Joists & Brick Facade
(2) Concrete Frame
(3) Post Tensioned Concrete Slabs, Sheer Walls, Window Wall Exteriors, and Drywall Partitions.

Average Square Foot Costs
2002

Commercial

Square Foot Costs do not include Land, Furnishings, and Financing.

	Division	Remodeling Office Interior	Office 1-Story (1)	Office 2-Story (2)	Office Up To 5-Story (3)	High Rise (City) Above 5-Story (4)	Parking Ramp (3)
1.	General Conditions	$ 2.15	$ 3.50	$ 3.50	5.00	5.50	3.90
2.	Site Work & Demolition	1.30	1.80	1.80	1.80	1.60	1.15
3.	Concrete	-	5.20	8.25	17.50	6.20	19.70
4.	Masonry	-	8.75	8.25	12.50	3.10	1.25
5.	Steel	-	-	7.50	2.70	1.95	1.35
6.	Carpentry and Millwork	5.40	2.90	2.70	4.25	4.65	1.00
7.	Moisture Protection	-	5.20	2.60	1.80	2.20	2.30
8.	Doors, Windows and Glass	1.45	5.40	5.40	11.20	20.80	1.10
9.	Finishes & Sheet Rock	9.25	9.70	9.65	11.40	12.20	1.50
10.	Specialties	-	1.40	1.40	1.45	1.45	-
11.	Equipment	-	4.35	3.85	3.20	4.30	-
12.	Furnishings	-	.90	1.10	1.20	1.30	-
13.	Special Construction	-	-	-	-	-	-
14.	Conveying	-	-	2.50	4.10	4.70	2.45
15.	Mechanical:						
	Plumbing	2.65	4.20	4.40	5.10	5.10	1.75
	Heating and Ventilation	1.40	5.30	5.00	6.20	6.60	-
	Air Conditioning	1.40	5.25	5.10	5.75	6.40	-
	Sprinklers	-	1.65	1.65	1.75	1.75	-
16.	Electrical	5.40	6.60	7.25	8.00	9.20	2.05
	SUB TOTAL	$30.40	$72.10	$81.90	$104.90	$ 99.00	$39.50
	CONTRACTOR'S SERVICES	3.05	7.20	8.20	10.80	9.90	4.00
	Construction Sq.Ft. Cost	$33.45	$79.30	$90.10	$115.70	$108.95	$43.50
	Add for Architect's Fee	3.45	6.20	7.00	8.20	8.50	3.20
	TOTAL Sq.Ft. Costs	$36.90	$85.50	$97.10	$123.90	$117.45	$46.70
	Deduct for Tenant Areas Unfinished	-	-	20.00	22.00	26.00	-
	Per Car Average	-	-	-	-	-	$12,300
	Add: Masonry Enclosed Ramp	-	-	-	-	-	4.60
	Add: Heated Ramp	-	-	-	-	-	4.40
	Add: Sprinklers for Closed Ramps	-	-	-	-	-	1.75
	Deduct for Precast Concrete Ramp	-	-	-	-	-	7.00

(1) Wall Bearing Masonry and Steel Joists and Decks - No Basement
(2) Wall Bearing Masonry and Precast Concrete Decks - No Basement
(3) Concrete Frame
(4) Steel Frame and Window Wall Facade

Average Square Foot Costs
2002

Commercial and Industrial

Square Foot Costs do not include Land, Furnishings, Landscaping, and Financing.
See Division 13 for Metal Buildings

	Division	Store 1-Story (1)	Store 2-Story (2)	Production 1-Story (1)	Warehouse 1-Story (1)	Warehouse and 1-Story Office (1)	Warehouse and 2-Story Office (3)
1.	General Conditions	$ 3.25	$ 3.25	$ 3.15	$ 2.80	$ 2.90	$ 2.60
2.	Site Work & Demolition	1.65	1.25	1.45	1.40	1.70	1.35
3.	Concrete	4.70	8.65	4.70	4.60	5.10	7.85
4.	Masonry	6.85	7.40	5.70	5.80	6.50	3.50
5.	Steel	7.60	4.00	7.60	7.35	8.00	6.10
6.	Carpentry	2.20	2.25	1.45	1.40	1.95	1.75
7.	Moisture Protection	3.80	2.15	4.10	3.90	3.90	1.95
8.	Doors, Windows and Glass	4.40	4.20	3.45	1.80	3.60	3.20
9.	Finishes	5.00	4.50	4.30	1.55	2.65	4.55
10.	Specialties	1.35	1.35	1.00	.95	.95	.95
11.	Equipment	1.30	1.45	1.35	-	.70	.85
12.	Furnishings	-	-	-	-	-	-
13.	Special Construction	-	-	-	-	-	-
14.	Conveying	-	3.00	-	-	-	-
15.	Mechanical:						
	Plumbing	3.30	3.00	3.50	1.95	2.50	2.30
	Heating and Ventilation	3.50	3.45	2.65	2.75	3.65	3.30
	Air Conditioning	3.75	3.85	2.50	-	.95	.85
	Sprinklers	1.65	1.70	1.55	1.60	1.65	1.60
16.	Electrical	6.00	5.75	6.55	3.85	4.00	3.60
	SUB TOTAL	$60.30	$61.20	$55.00	$41.70	$50.70	$46.30
	CONTRACTOR'S SERVICES	6.00	6.10	5.50	4.20	5.10	4.60
	Construction Sq.Ft. Cost	$66.30	$67.30	$60.50	$45.90	$55.80	$50.90
	Add for Architect's Fees	4.80	4.30	4.30	3.30	3.85	3.60
	TOTAL Sq.Ft. Cost	$71.10	$71.60	$64.80	$49.20	$59.65	$54.50

(1)   Wall Bearing Masonry and Steel Joists and Decks - Brick Fronts
(2)   Wall Bearing Masonry and Precast Concrete Decks - Brick Fronts
(3)   Precast Concrete Wall Panels, Steel Joists, and Deck

Average Square Foot Costs
2002

Educational and Religious

Square Foot Costs do not include Land, Furnishings, Landscaping and Financing.

Division	Elementary School 1-Story (1)	High School 2-Story (1)	Vocational School 3-Story (1)	College Multi-Story (2)	Gym (3)	Dorm 3-Story (1)	Church (3)
1. General Conditions	$ 5.50	$ 5.40	$ 5.45	$ 5.35	$ 4.20	$ 4.90	5.30
2. Site Work	2.90	3.10	2.90	2.90	1.45	1.90	2.75
3. Concrete	4.80	5.00	3.60	17.30	5.00	3.80	4.55
4. Masonry	11.00	10.40	10.80	11.20	11.00	9.70	22.45
5. Steel	7.60	9.30	9.40	3.15	9.40	8.00	1.40
6. Carpentry	3.20	3.05	2.75	1.95	1.75	2.70	5.90
7. Moisture Protection	4.20	2.90	3.30	3.35	4.50	3.90	4.65
8. Doors-Windows-Glass	5.30	5.20	5.50	6.90	4.90	5.45	6.40
9. Finishes	8.70	7.25	8.10	7.25	6.15	8.45	8.80
10. Specialties	1.40	1.35	1.35	1.70	1.35	1.35	1.00
11. Equipment	3.85	3.70	5.55	4.85	7.30	4.75	6.60
12. Furnishings	-	.75	.95	1.00	-	.95	3.85
13. Special Construction	-	-	-	-	-	-	-
14. Conveying	-	-	2.75	2.75	-	2.75	-
15. Mechanical:							
Plumbing	6.50	6.65	6.65	6.10	4.30	6.10	4.20
Heating-Ventilation	9.90	9.60	10.90	9.80	6.30	6.70	6.85
Air Conditioning	-	-	-	-	-	-	5.30
Sprinklers	1.70	1.70	1.70	1.70	1.70	1.70	-
16. Electrical	9.45	10.15	10.75	10.85	6.70	7.80	6.80
SUB TOTAL	$86.00	$85.50	$92.40	$98.10	$76.00	$80.80	96.80
CONTRACTOR'S SERVICES	8.60	8.60	9.30	9.80	7.60	8.10	9.70
Constr. Sq.Ft. Cost	$94.60	$94.10	$101.70	$107.90	$83.60	$88.90	106.50
Add Architect's Fees	6.80	6.80	7.20	7.70	6.00	6.40	7.50
TOTAL Sq.Ft. Cost	$101.40	$100.90	$108.90	$115.60	$89.60	$95.30	$114.00

(1) Wall Bearing Masonry and Steel Joists and Decks
(2) Concrete Frame and Masonry Facade
(3) Masonry Walls and Wood Roof

Average Square Foot Costs
2002

Medical and Institutional

Square Foot Costs do not include Land, Landscaping, Furnishings, and Financing.

		Clinic 1-Story (1)	Doctors Office 1-Story (1)	Hospital 1-Story (1)	Hospital Multi-Story (2)	Nursing Home 1-Story (1)	Housing for Elderly 1-Story (1)
1.	General Conditions	$ 5.80	$ 5.85	$ 7.85	$ 7.85	$ 7.10	$ 6.50
2.	Site Work	1.80	1.80	2.75	2.85	3.00	3.05
3.	Concrete	5.85	5.80	6.00	19.00	6.20	6.25
4.	Masonry	10.20	9.20	16.60	11.60	11.80	10.50
5.	Steel	6.65	6.60	8.10	1.80	1.85	1.55
6.	Carpentry	5.50	5.40	4.70	3.10	3.25	3.05
7.	Moisture Protection	4.15	4.05	4.40	1.85	4.35	4.40
8.	Doors, Windows and Glass	6.60	7.00	8.80	5.65	7.65	6.15
9.	Finishes	8.80	8.60	11.10	10.80	10.80	8.90
10.	Specialties	1.25	1.25	1.35	1.30	1.30	1.25
11.	Equipment	6.70	5.40	10.30	10.10	5.10	2.20
12.	Furnishings	1.00	1.00	1.05	1.15	.95	2.00
13.	Special Construction	.70	.70	1.50	1.50	.70	.70
14.	Conveying	-	-	-	5.60	-	5.20
15.	Mechanical:						
	Plumbing	8.80	9.40	14.50	13.80	9.45	9.10
	Heating and Ventilation	8.00	7.15	10.50	10.80	5.50	4.35
	Air Conditioning	6.20	5.90	10.80	9.35	6.20	2.35
	Sprinklers	1.70	1.70	1.70	1.70	1.90	1.70
16.	Electrical	11.50	9.80	18.00	16.20	10.50	9.20
	SUB TOTAL	$101.20	$96.60	$140.00	$136.00	$97.60	$88.40
	CONTRACTOR'S SERVICES	10.10	9.70	14.00	14.00	9.80	8.80
	Construction Sq.Ft. Cost	$111.30	$106.30	$154.50	$150.00	$107.40	$97.20
	Add for Architect's Fees	8.00	7.70	11.00	11.00	7.60	7.00
	TOTAL Sq.Ft. Cost	$119.30	$114.50	$165.00	$161.00	$115.00	$104.20

| Average Cost per Bed | | | | $107,000 | $106,000 | $44,000 | |
| Average Cost per Patient Room | | | | $155,000 | $150,000 | $74,000 | |

Average Cost to Remodel Approximately 50% Cost of New

(1)   Wall Bearing Masonry and Steel Joists and Decks
(2)   Concrete Frame and Masonry Facade

Average Square Foot Costs
2002

Government and Finance

Square Foot Costs do <u>not</u> include Land, Landscaping, Furnishings, and Financing.

		Bank & Drive-Thru 1-Story (1)	Bank & Insur. Bldg High-Rise (3)	Govt. Office Bldg Multi-Story (2)	Post Office 1-Story (1)	Court House 3-Story (2)	Jail and Prison (2)
1.	General Conditions	$ 6.40	$ 7.35	$ 7.35	$ 6.00	$ 7.25	$ 8.00
2.	Site Work	3.40	3.10	2.60	2.15	2.35	3.10
3.	Concrete	8.00	19.50	20.50	8.80	17.50	24.20
4.	Masonry	12.10	8.35	13.90	8.35	13.75	15.40
5.	Steel	3.60	1.65	2.00	2.75	1.70	4.05
6.	Carpentry	3.60	3.25	3.35	2.90	2.10	3.75
7.	Moisture Protection	3.30	3.45	3.80	3.00	3.85	4.40
8.	Doors, Windows and Glass	8.50	20.00	8.80	6.75	6.30	8.15
9.	Finishes	12.20	13.50	14.20	12.10	11.45	9.80
10.	Specialties	1.35	1.35	2.60	1.40	2.20	5.40
11.	Equipment	10.50	7.00	7.00	9.30	8.80	10.40
12.	Furnishings	1.20	1.10	-	1.65	-	-
13.	Special Construction	-	-	-	-	-	-
14.	Conveying	-	2.25	2.90	1.50	2.10	2.90
15.	Mechanical						
	Plumbing	4.90	6.20	6.05	5.60	8.40	14.00
	Heating and Ventilation	4.80	6.00	6.50	4.75	6.95	12.40
	Air Conditioning	3.95	4.40	5.55	5.60	9.70	10.35
	Sprinklers	1.75	1.70	1.70	1.70	1.70	1.70
16.	Electrical	10.50	12.60	13.20	10.00	12.50	20.40
	SUB TOTAL	$100.05	$122.75	$122.00	$94.30	$ 118.60	$158.40
	CONTRACTOR'S SERVICES	10.00	12.30	12.20	9.40	11.80	15.80
	Construction Sq.Ft. Cost	$110.05	$135.05	$134.20	$103.70	$130.40	$174.20
	Add for Architect's Fees	8.00	10.20	10.20	7.50	9.30	12.40
	TOTAL Sq.Ft. Cost	$118.05	$145.25	$144.40	$111.20	$139.70	$186.60

(1) Wall Bearing Masonry and Precast Concrete Decks
(2) Concrete Frame and Masonry Facade
(3) Concrete Frame and Window Wall

Construction Industry Statistical Data
<u>Number of Employees in Contract Construction Industry</u>
And Total U.S. Labor Force 2001 (2000 Statistics)

1. Total Population Employment Status

	Total 2000 Employees[1]
Mining	557,000
Contract Construction (Breakdown below)	6,840,000
Manufacturing	17,939,000
Transportation and Public Utilities	7,096,000
Wholesale Trade	7,066,000
Retail Trade	23,331,000
Finance, Insurance and Real Estate	7,716,000
Service and Miscellaneous	40,644,000
Government - Federal	2,614,000
State	4,807,000
Local	13,217,000
Total Non-Agricultural	**132,170,000**
Agriculture	3,163,000
Total Employed	**135,333,000**
Unemployed	3,326,000
Total Labor Force	**138,659,000**

2. Contract Construction Employment Status 2000

	Total Construction[2] Employment	Workers	Architects Engineers Managers Supervision Clerical
General Building Contractors			
Residential - Contractors	791,300	510,300	281,000
Operative - Developers	30,600	12,500	18,100
Non Residential - Contractors	654,800	468,700	186,100
Total General Building	**1,476,700**	**991,500**	**485,200**
Special Trade Contractors			
Plumbing, Heating & Air Conditions	934,900	688,000	246,100
Painting and Decorating	217,900	175,900	42,000
Electrical Work	863,300	677,100	186,200
Masonry, Plastering, Tile and Stone	543,200	469,800	73,400
Carpentry and Flooring	314,900	234,900	80,000
Roofing and Sheet Metal	237,400	179,700	57,700
Other Miscellaneous	1,071,000	781,800	289,300
Total Special Trade	**4,182,600**	**3,207,000**	**974,700**
Heavy Construction			
Highway and Street	228,000	177,000	51,000
Other Heavy	590,700	498,200	92,500
Total Heavy Construction	**818,700**	**675,200**	**143,500**
Total Construction Employees	**6,478,000**	**4,873,900**	**1,603,400**

---

(1) *Employment and Earnings May 2001* - U.S. Department of Labor, Bureau of Labor Statistics
(2) AC&E Projections from 1998 Construction Review which ceased publication in 1998.

Construction Industry Statistical Data
2001 (2000 Statistics)
Gross Earnings and Hours

| | United States[1] | | | Metropolitan Areas[2] | |
	Average Hourly Earnings	Average Weekly Earnings	Gross Hours	Hourly Earnings	40-Hour Weekly Earnings
Mining	17.35	792.90	45.7		
Contract Construction	18.24	693.12	38.0	28.10	1,124.00
Manufacturing	14.79	585.68	39.8		
Transportation and Public Utilities	16.69	624.47	38.6		
Wholesale Trade	15.81	584.43	38.6		
Retail Trade	9.77	282.35	28.9		
Finance, Insurance and Real Estate	15.85	584.87	36.9		
Services	14.57	478.44	32.7		
Agriculture					
Average	13.74	474.03	34.5		
Contract Construction					
General Building Contractors	18.29	702.34[2]	37.8	26.80[3]	1072.00
Residential	16.72	613.62	36.7		
Operative	19.53	720.66	36.9		
Non Residential	18.88	740.10	39.2		
Special Trade Contractors					
Plumbing, Heating and Air Cond.	19.20	748.88	38.0	32.00	1,280.00
Painting and Decorating	16.18	580.88	38.9	27.80	1,112.00
Electrical Work	20.01	820.28	35.9	33.20	1,328.00
Masonry, Plastic, Tile and Stone	18.12	652.32	36.6	29.60	1,184.00
Roofing and Sheet Metal	15.91	528.08	36.6	26.00	1,040.00
Carpentry and Flooring	18.52	675.98	33.1	29.00	1,160.00
Heavy Construction					
Highway and Street	17.12	679.66	40.9		
Other Heavy	17.45	719.88	41.3		
Combined Construction					
Average (39.1 hours)	18.29	702.34	38.4	28.10	1,098.00
Average (40.0 hours)				28.10	1,124.00
[2] Skilled 80%				29.50	1,180.00
Laborers 20%				22.50	900.00
Weighted Skilled and Laborers				28.10	1,124.00

[1] *Employment and Earnings May 2001 -*
U.S. Department of Labor, Bureau of Labor Statistics

[2] A.C.& E. 60 Metropolitan Areas - Weighted Index
Skilled Average = 28.00
  Plumbers & Sheet Metal - 16% @ 32.00
  Carpenters - 36% @ 29.00
  Cement Masons - 2% @ 27.10
  Iron Workers & Op.Engrs. - 8% @ 31.50

  Brick Layers - 7% @ 29.60
  Painters - 5% @ 27.80
  Roofers - 5% @ 26.10
  Electricians - 12% @ 33.20
  10 Misc. Trades - 9% @ 28.00

[3] 2 Carpenters @ 29.00 to 1 Laborer @ 22.50 = 26.80

Construction Industry Statistical Data
2001 (2000 Statistics)
New Construction Put In Place in the United States
(Millions of Dollars)

	PRIVATE[1]		PUBLIC[2]
Residential			
New Housekeeping			
1 Unit	230,000		
2 or More Units	28,500		
Total New Housekeeping		258,500	
Addition & Alterations (Improvements)		63,500	
Total Residential		**322,000**	
Buildings			
Industrial		30,100	2,000
Commercial, Office		54,700	-
Commercial, Other		59,700	-
Religious		7,900	-
Educational		10,600	27,000
Hospital and Institutional		13,300	5,100
Hotels and Motels		17,100	-
Miscellaneous Buildings		10,100	27,000
Housing and Redevelopment		-	5,300
Total Buildings		**203,400**	**59,070**
Heavy Construction			
Public Utilities			
Tele Communications			
Railroads			
Electric Light and Power			
Gas			
Petroleum Pipe Line			
Total Public Utilities			
Sewer			12,300
Water			6,800
Miscellaneous			15,600
Total Heavy Construction			**34,700**
Highways and Streets			115,200
Miscellaneous			
Farm	2,363		-
Military	-		3,700
Conservation	-		8,000
Total Miscellaneous		**2,363**	**11,700**
GRAND TOTAL		**574,000**	**158,000**

TOTAL DOLLAR VALUE
TOTAL PRIVATE & PUBLIC CONSTRUCTION 1999   705,000,000,000
TOTAL PRIVATE & PUBLIC CONSTRUCTION 1998   642,000,000,000
1998 INCREASE   63,000,000,000

PHYSICAL VOLUME INCREASE   +7.0%

[1] U.S. Department of Commerce
[2] *Not available since Construction Review ceased publication in 1998.* 2000 Statistics Estimated.

Construction Industry Statistical Data
2001 (2000 Statistics)
Number of Construction Workers by Trade Designation

Workers (Tradesmen)	Number of Employees [1]	Percent of Workers [2]
Basic Trades		
Bricklayers	257,400	5.0 %
Carpenters	1,362,400	26.6
Cement Masons	113,250	2.2
Iron Workers	133,800	2.6
Laborers	1,029,600	20.0
Operating Engineers	164,700	3.2
Truck Drivers	30,900	.6
Specialty Trades		
Carpet and Resilient Floor Layers	164,700	3.2
Elevator Constructors	20,600	.4
Glaziers	25,700	.5
Lathers	10,300	.2
Painters	175,900	3.7
Plasterers	20,600	.4
Roofers and Sheet Metal	179,700	4.3
Terrazzo Workers	10,300	.2
Tile Setters	46,300	.9
Acoustical Tile Setters	36,300	.7
Mechanical and Electrical Trades		
Electricians	677,100	11.4
Plumbers and Pipe Fitters	688,800	11.9
Sheet Metal Workers	100,300	2.0
Total Workers	**5,248,650**	**100.00%**

Architects, Engineers and Contractors		Percent of Total
Architects and Engineers	544,000	38.00
Construction Managers and Estimators	576,000	40.00
Accountants and Clerical	320,000	22.00
Total	**1,440,000**	**100.00%**

All Workers Total	**6,688,650**	

[1] *Construction Review 1998* - U.S. Department of Commerce & AC&E Projections for 2000 Statistics
[2] Percentages - AC&E

Note: Metro Weighted Wage Rate Averages and Modifiers computed from this table of approximate number of active Tradesmen in the industry.

U.S. Measurements & Metric Equivalents - (For Construction)

Weights & Measures (U.S. System)

Length	1 Mile = 320 Rods = 1,760 Yards = 5,280 Feet
Area	1 Mile = 640 Acres = 102,400 Sq. Rods = 3,097,600 Sq. Yards
Area	1 Acre = 160 Sq. Rods = 4,840 Sq. Yards = 43,560 Sq. Feet
Volume	1 Cu. Yard = 27 Cu. Feet = .211 Cord
Capacity	1 Gallon = 4 Quarts = 8 Pints = .13337 Cu. Feet
Weight & Mass	1 Ton = 2,000 Pounds = 32,000 Ounces

U.S. to Metric - Area, Length, Volume & Mass Conversion Factors

Quantity	From U.S.	To Metric	Multiply By
Length	mile	km	1.609 344
	yard	m	0.914 4
	foot	m	0.304 8
	inch	mm	25.4
Area	square mile	km^2	2.590 00
	acre	m^2	4 046.87
	square yard	m^2	0.836 127
	square foot	m^2	0.092 903
	square inch	mm^2	645.16
Volume	cubic yard	m^3	0.764 555
	cubic foot	m^3	0.028 316 85
Mass-(Weight)	lb	kg	0.453
Mass-(Length)	plf	kg/m	1.488
Mass-(Square)	psf	kg/m^2	4.882
Mass-(Cube)	pcf	kg/m^3	16.018

Length Conversion Tables from Fractions to Decimals to Millimeters

Fraction	Decimal	Millimeters	Fraction	Decimal	Millimeters
1/16"	.0625	1.588	9/16"	.5625	14.288
1/8"	.125	3.175	5/8"	.625	15.875
3/16"	.1875	4.763	11/16"	.6874	17.463
1/4"	.250	6.350	3/4"	.750	19.050
5/16"	.3125	7.938	13/16"	.8125	20.638
3/8"	.375	9.525	7/8"	.875	22.225
7/16"	.437	11.113	15/16"	.9375	23.813
1/2"	.500	12.700	1"	1.000	25.400

Weights - Construction Products

Substance	Wt.Lb. per Cu.Ft.	Substance	Wt.Lb. per Cu.Ft.	Substance	Wt.Lb. per Cu.Ft.
Aluminum	165	Lead	710	Concrete	144
Brass	534	Paper	58	Cement	90
Bronze	509	Glass	156	Clay	100
Copper	556	Granite	165	Sand	100
Iron	450	Limestone	160	Asphalt	81
Steel	490	Brick	140	Wood	40

Computation of Area

Square or Rectangular	=	Length x Width or Height
Triangle	=	Base x ½ Height
Circumference of Circle	=	Diameter x 3.1416
Area of Circle	=	Radius Squared x 3.1416
Cube of a Column or Cylinder	=	Radius Squared x 3.1416 x Depth
Cube of a Solid	=	Length x Width x Depth
Area of Ellipse	=	½ Length x ½ Height x 3.1416

UNITED STATES INFLATION RATES
July 1, 1970 to June 30, 2001
AC&E PUBLISHING CO. INDEX

Start July 1

YEAR	AVERAGE %	LABOR %	MATERIAL %	CUMULATIVE %
1970-71	8	8	5	0
1971-72	11	15	9	11
1972-73	7	5	10	18
1973-74	13	9	17	31
1974-75	13	10	16	44
1975-76	8	7	15	52
1976-77	10	5	15	62
1977-78	10	5	13	72
1978-79	7	6	8	79
1979-81	10	5	13	89
1980-81	10	11	9	99
1981-82	10	11	10	109
1982-83	5	9	3	114
1983-84	4	5	3	118
1984-85	3	3	3	121
1985-86	3	2	3	124
1986-87	3	3	3	127
1987-88	3	3	4	130
1988-89	4	3	4	134
1989-90	3	3	3	137
1990-91	3	3	4	140
1991-92	4	4	4	144
1992-93	4	3	5	148
1993-94	4	3	5	152
1994-95	4	3	6	156
1995-96	4	3	6	160
1996-97	3	3	3	163
1997-98	4	3	4	167
1998-99	3	3	2	170
1999-2000	3	2	3	173
2000-2001	4	4	4	177
2001-2002	3	4	3	180

A

Abrasive - nosings 302.35
Access doors 1001
Access floors 1305
Accordion doors............................... 1020
Acoustical tile.................................... 905
Admixtures - concrete..................... 301.32
    - masonry.................................. 409.5
Automotive equipment 1115
Air conditioning 1507
Altar.. 1111
Anchor - bolts 502.1
    - slots 302.32
Anechoic chambers 1316.1
Antenna systems 1607.1
Appliances - residential..................... 1106
Ashlar stone 407.2
Asphalt - expansion joint................... 301.35
    - floor tile 908.1
    - pavement.............................. 206.1
    - shingles 707.1
Astronomy observation domes 1315
Athletic equipment - indoor 1107
    - outdoor 208.1
Audiometric rooms............................ 1316.2
Auger holes 202.2
Automatic door openers..................... 808.3

B

Backfill... 202.3
Backstops - indoor basketball............. 1107.1
    - outdoor basketball 208.2
Bank equipment................................ 1113
Bar joists 504
Base - ceramic tile 903.1
    - concrete................................. 301.22
    - quarry tile 903.2
    - resilient 908.5
    - terrazzo 904.1
    - wood..................................... 603.4
Baseboard heating - electric 1608.2
    - hot water................................ 1508.5
Bathroom accessories 1026
Bathtub enclosures 1027
Beams - formwork 302.14
    - grade 302.12
    - laminated 604
    - precast.................................. 305.10
    - steel..................................... 501.1
    - wood..................................... 604.2
Benches - exterior 208.3
Blackboards..................................... 1002

Bleachers - exterior.......................... 208.3
    - interiors 1107.2
Block - concrete 402
    - glass 404
    - gypsum 4405
Blocking - carpentry 601.31
Boards - chalk.................................. 1002
    - directory 1006
    - dock 1008
Boilers.. 1506.4
Booths - parking............................... 1115.3
    - telephone 1025
Bowling alleys 1302
Box - mail 1018
Brick - masonry................................ 401
Bridging... 601.13
Bucks - doors................................... 601.31
Building paper 601.34
Built-up roofing................................. 705.1
Bulletin boards 1006
Bumpers - dock................................ 1008
Burglar alarm system 1607.3
Bus duct.. 1601
Bush hammer concrete...................... 301.25

C

Cabinets - hospital 1101
    - kitchen 1103
    - laboratory.............................. 1104.2
    - school 1102
    - shower 1027
Cable - tray 1601.9
Caissons... 203.4
Cants - roof 601.33
Carborundum 301.31
Carpentry - finish 602
    - rough 601
Carpet... 909
Castings.. 502.2
Cast iron - manhole covers 205.3
Catch basins 205.3
Caulking.. 710
Cellular deck - metal 505.3
Cement - colors 301.33
    - grout 301.36
    - Portland 409.1
Cementitious decks........................... 307
Ceramic tile...................................... 903.1
Chain link fence 207.4
Chairs 1205.2
Chalkboard 1002
Chamfer strip 302.33

INDEX

Chiller - water 1507.3
Chimney - concrete 1303.1
   - metal.................................... 1303.2
   - radial brick 1303.1
Church equipment 1111
Chutes - linen & rubbish 1003
Circuit breakers 1604.5
Clay facing tile 403
Clean rooms 1304
Clear and grub............................. 201.2
Clock & program system................ 1607.4
Coat & hat racks 1004
Coiling partitions - metal
           & wood.................. 1020
Columns - forms 302.13
Compaction - soil......................... 202.4
Compactors - waste...................... 1028
Concrete - prestressed 305
Condensing units......................... 1507.2
Conduit - electrical....................... 1601.1
Confessionals............................. 1111.2
Cooling towers............................ 1507.4
Core drilling 201.3
Cork - flooring 908.2
   - tackboard............................. 1002
Corner guards - steel.................... 1029
Counter tops............................... 603.2
Courts - tennis 208.1
Cranes....................................... 1406
Cribbing 204.6
Culverts 205.2
Curbs - asphalt & concrete 206.2
Curing - concrete 301.34
Curtain - walls............................. 809
   - cubicles & tracks..................... 1005

D

Dampproofing.............................. 702
Darkroom - prefab 1004.2
Deck - laminated wood 604.3
   - metal.................................... 505
Demolition 201
Dental equipment 1104.17
Detention equipment..................... 1114
Dewatering 205.5
Dimmer switch 1601.5
Directory boards 1006
Display & Trophy cases 1007
Disposal excavation...................... 202.5
Dock equipment - loading 1116
Doctors register 15
Domes - skyroof 709.2

Door - blast................................ 804.1
   - frames, metal......................... 804.1
   - frames, wood 801
   - hardware............................... 803
   - flexible.................................. 804.3
   - folding 1020
   - glass, all................................ 804.4
   - hangers................................. 804.5
   - industrial............................... 804.15
   - metal.................................... 801
   - overhead............................... 804.7
   - plastic laminate 804.8
   - revolving 804.9
   - rolling 804.10
   - shower 804.11
   - sliding................................... 804.12
   - traffic 804.14
   - vault 804.16
   - weather stripping..................... 811
   - wood 803
Dovetail slot 302.32
Drywall - gypsum 602.24
                 and 902
Ducts - electrical 1601.6
Dumbwaiters............................... 1403

E

Ecclesiastical equipment................ 1111
Elastomeric roofing 706
Elevators.................................... 1401
Epoxy flooring 301.25
Escalators.................................. 1402
Evaporators................................ 1507.6
Excavation.................................. 202.2
Exit lights 1605.1
Expansion joints - concrete............. 301.35
   - covers 1009

F

Fertilizing 209.2
Film viewer, X-ray 1104.10
Finish - hardware 810
Finishing - concrete...................... 301.2
Fire alarm system 1607.2
Fire - brick................................. 401.4
   - doors.................................... 804.6
   - escapes 502.4
   - extinguishers.......................... 1010
   - hydrants 1505.3
Fireplace accessories 412.4
Fixtures - electrical...................... 1605.1
   - plumbing 1504.4

Flagpoles 1011
Flagstone 407.3
Flashing 708.3
Flue lining 406.2
Folding doors & partitions 1020
Folding gates 1014
Food service equipment 1108
Fuel tanks 1501.8
Fuel handling equipment 1506.1
Fume hoods 1104.4

G

Generators - emergency 1602.1
Glass and glazing 812
Glazed - block.................................... 402.7
    - brick 401.2
Glue laminated 604
Grading - site 202.1
Grandstands 1306.1
Granite - building 407.13
Greenhouses 1307
Grilles & louvers 1016
Grounding.. 1604.2
Grounds.. 601.32
Guard rails 207.4
Gunite... 304.4
Gutters & downspouts 708.1
Gymnastics equipment 1107.7
Gypsum - deck 307.2
    - drywall 602.24
                                                  and 902
    - lath 901.12

H

Hand excavation............................... 202.2
Hardboard 602.13
Hardeners - concrete........................ 301.37
Hardware - drapery........................... 1204
    - finish 810
    - rough 607
Hatches - floors & roof 1012
Heating 1506
Hollow metal doors 801
Hollow metal toilet partitions 1027
Hospital equipment........................... 1104
Incinerators...................................... 1308
Insulated rooms 1309
Insulation - building 703
- pipe & duct 1501.9
Integrated ceiling 1509.4
Intercom systems 1607.5

J

Joints - control & expansion 301.35
Joists - steel.................................... 504
    - wood 601.12

K

Kitchen - equipment 1108

L

Laboratory equipment 1104.2
Landscaping 209
Lathing 901.1
Laundry chutes 1003
Laundry equipment 1109
Lavatories 1504.4
Lead - shielding............................... 1104.10
Letter & signs 1022
Lift slab 304.1
Lifts - garage................................... 1404.2
    - stage.. 1404.3
    - man 1404.4
Light gage framing 507
Lighting 1605
Lightning protection.......................... 1606
Limestone 407.11
Linen chutes 1003
Lintels - steel.................................. 502.10
Loading dock equipment 1116
Lockers 1015
Locksets 810
Louvers.. 1016

M

Mail chutes & boxes......................... 1018
Manholes 205.3
Map rail 1002
Marble.. 407.12
Marble veneer 906.2
Masonry - restoration 414
Medical equipment............................ 1104
Medicine cabinets 1027
Mercury vapor lights......................... 1605.2
Mesh - partitions 1017
Metal - buildings.............................. 1310
    - lath.. 901.11
    - partitions 1019
Metering- mechanical........................ 1502.6
    - electrical................................... 1604.1
Mirrors 812.12
Miscellaneous metals........................ 502
Monorails 1406
Mortar 409

Movable partitions 1019
Moving buildings 201.1
Moving stairs & walks 1402

N

Nosings - stairs 302.34

O

Oil storage tanks 1501.8
Open web joists 504
Ornamental metals 503

P

Painting - building 912
Pan forms 302.15
   - stairs 502.12
Paneling 502.23
Panels - asbestos cement 809.3
   - concrete 306.3
Parget flooring 910.2
Parking equipment 1115
Partitions - acoustical 1019
   - drywall 902
   - movable 1019
   - toilet 1027
Pavers - floor 906.4
Pedestal floors 1304
Pegboard 602.22
Pews - church 1111.3
Piling - bearing 203
   - sheet 204.1
Pipe & fittings 1501.1
Placing concrete 301.1
Plaques 1022
Plaster 901.2
Plastic - pipe 1501.1
   - skylights 709.1
Plumbing 1502
Pneumatic tube systems 1405
Poles - electrical 1601
Polyethylene film 301.39
Pool equipment 1504.5
Post-tension concrete 304.2
Powered scaffolding 1408
Precast - concrete 306
   - panels 408
Pre-engineered buildings 1310
Prestressed concrete 305
Prison equipment 1114
Projection screens 1203
Protected metal 506
Public address & paging system 1607.6

Pumps - condensate 1501.5
   - heat 1501.5
   - water 1504
   - well 1502
   -wellpoint 205.2

Q

Quarry tile 903.2

R

Radiation protection 1311
Rafters - wood 601.14
Railroad work 212
Ranges - shooting 1107.8
Receptors - shower 1021
Refrigeration 1507
Reinforcing - concrete 303.1
Residential - appliances 1106
Resilient flooring 908
Restoration - masonry 4
Rip-rap 204.5
Roads & walks 206
Roof decks - insulation 704
   - metal 505
Roofing - built up 705.1
   - skylights 709.1
   - traffic 706
   - ventilators 709
Rubber - flooring 908.4
   -roofing 706
Rubbish chutes 1003
Running tracks 208.1

S

Sandblasting - walls 414.2
Saunas 1312
Scaffolding 416
Scales 1313
School equipment 1105
Scoreboards 1107.3
Screeds 301.2
Screens - projection 1203
   - urinal 1027
   - window 808.1
Sealants 710
Seamless flooring 907.1
Seating 1207
Seeding & sodding 209.1
Septic tanks 1503.4
Service station equipment 1117
Sewage treatment 1503.4
Shades 1202
Sheathing 601.16

Sheetrock 902
Shelters - loading dock 1116.4
Shelving - library 1110
    - storage 1023
Shingles - asphalt 707.1
    - asbestos 707.2
    - wood 701.3
    - slate 701.5
    - tile 701.6
Shoring 302.2
Siding - metal 602
    - wood 602.1
Signs 302.2
Site drainage 20.5
Skydomes 1315
Slate - chalkboard 1002
    - panels 906.3
    - roofing 707.6
Slides - playground 208.2
Snow melting equipment 1608.1
Soffits 602.17
Soil - compaction 202.4
    - poisoning 210.1
    - testing 202.6
Soundproof enclosures 1025
Sound insulated room 1316
Sound & vibration control 1317
Spiral stairs 502.13
Sprinklers - building 1505.1
Stage & theatrical equipment 1112
Steel - buildings 1310
    - studs 902.4
Sterilizers 1104.9
Stone 407
Stone - flagging & paving 906.12
Stone veneer 906
Structural steel 501
Stucco 901.24
Studs - metal wall 902.4
    - wood 601.11
Substations 1603.1
Sun control devices 1024
Swimming pools 1318
Swing - playground 208.12
Switchgear 1601.4
Switchboards 1604.4

T

Tackboards 1002
Tanks - oil, gasoline & water 1501.8
Telephone booths 1025
Telephone equipment 1607.1
Television antennae 1607.7
Temperature control equipment 1609.3
Tennis courts 208.1
Termite - soil treatment 210.1
Terra cotta 408.2
Terrazzo 904

Ties - form 302.31
Tile - acoustical 905
    - ceramic 90
    - plastic 913.2
    - quarry 903.2
    - resilient 908
    - roofing 707.6
Tilt up construction 304.3
Timber - framing 601.2
Toilet - accessories 1026
    - partitions 1027
Tops & bowls 1104.3
Tracks - running 208.1
Transformers 1604.6
Trash chutes 1003
Trees & shrubs 209.5
Trench covers 502
Truck - docks 1404.1
    - scales 1313
Trusses - wood 605.1

U

Underpinning 204.4
Utilities - sitework 205

V

Valves 1501.4
Venetian blinds 1202
Vibroflotation 210.3
Vinyl asbestos flooring 908.6
Vinyl flooring 908.5

W

Walks 206.3
Wall coverings 913
Wall urn 1030
Waterproofing 701
Water stops 302.35
Water treatment 1502.4
Windows 805
Wire - electric 1601.3
    - mesh 303.2
Wood - doors 803
    - flooring 910
    - framing 601.1
    - piling 204.2
    - treatment 606
    - trusses 605.1

X

X-ray equipment 1104.1

Notes

Notes

Notes

Notes

Notes

Notes

Notes

Notes